THOMAS CRAPPER

The Real Crapper Story

Richard E Doughty

Grosvenor House
Publishing Limited

This book is published by
Grosvenor House Publishing Ltd
28-30 High Street, Guildford, Surrey, GU1 3EL.
www.grosvenorhousepublishing.co.uk

A CIP record for this book
is available from the British Library

ISBN 978-1-78148-319-0

This book is dedicated in memory of my
Mother, Father & Grandparents

CONTENTS

Photograph courtesy of Thomas Crapper & Company Ltd.

INTRODUCTION

Over a period of fifty years Thomas Crapper nurtured a coveted reputation for service, innovation and excellence. Through sheer hard work he built an enviable, instantly recognisable iconic brand name, Thomas Crapper & Co, which continues to this day.

The Yorkshire lad from Waterside settled in London at an early age and progressed to become the most famous Plumber and Water Closet Maker in the world. Thomas Crapper was an outstanding Victorian Sanitary Engineer, Inventor and Entrepreneur. Thomas Crapper & Co was granted no fewer than six Royal Warrants including the ultimate seal of approval 'By Appointment to His Majesty The King'.

Thomas Crapper held a combined total of seventeen patents and registered designs, he successfully developed, improved and promoted 'Crapper's Silent Valveless Water Waste Preventer', the direct ancestor of the efficient leak-free British syphon that is still incorporated within Thomas Crapper & Company's water closet flushing cisterns today.

Some historians have mistakenly attributed '*Mr Thomas Crapper's new invention 'the flush toilet' being responsible for the abominable stench or 'great stink' of the River Thames, London during the 1850s and 1860s.*' In 1858 other academic opinion made reference to '*those pestiferous inventions – Water Closets, their only use being an accommodation to the lazy habits of mankind.*' Modern sanitation, in particular the invention of the water closet, is now acknowledged as one of the greatest advances in the prevention and spread of disease amongst the world's growing population.

Thomas Crapper is not known to have written any technical books or given lectures on sanitation, unlike fellow distinguished sanitary engineers S Stevens Hellyer, P J Davies or W P Buchan. Neither is he known to have voiced his opinion on sanitary issues of the day in the trade press unlike George Jennings. Thomas Crapper was a highly respected artisan with natural leadership ability to motivate, delegate and oversee. Once successful he focused his attention to developing improved sanitation designs and patents based on his own extensive practical experience.

Crapper's Improved
Registered Ornamental Flush=down W.C.

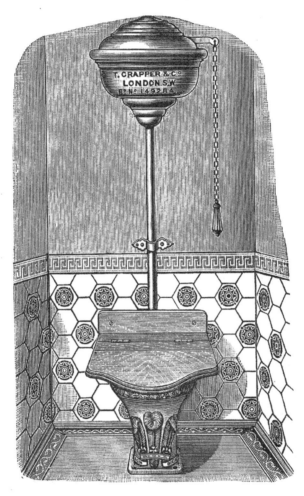

New Design

Water Waste

Preventer.

Reversible

Action.

No Brackets

required.

No Unions

or Joints

in Sight.

No. 165.

No. 165.   Improved Ornamental Flush-down W.C. Basin (Registered No. 145,823), Polished Mahogany Seat with Flap, New Pattern 3-gallon Cast-iron Syphon Cistern (Registered No. 149,284), and Pendant Pull, complete as shewn £5 12 6

Surnames have evolved over the course of history, however one name in particular, when mentioned, is guaranteed to evoke a reaction, the name 'Crapper'. The word Crapper is inextricably associated with the famous historical figure Thomas Crapper.

'Crapper' is often used in humour originating from an Americanism. During the First World War many American troops were stationed in Great Britain where they observed the brand name 'Crapper' displayed prominently on water closets and flushing cisterns. This caused some amusement that spread across the Atlantic and back again. The word Crapper, etymologists' conclude, is synonymous with Thomas Crapper, Sanitary Engineer, Inventor and Entrepreneur.

Extensive research has revealed factual errors within Wallace Reyburn's original, amusing biography of Thomas Crapper, whilst other publications, along with many websites, often include substantial embellishment. This book, *The Real Crapper Story* includes previously unpublished material about the life and times of Thomas Crapper e.g.

➢ Thomas belonged to a dynasty of seventeen plumbers and sanitary engineers including four Thomas Crappers.
➢ Thomas was not the first member of his family to work at Sandringham.
➢ The plight of two of Thomas Crapper's close relatives.
➢ The Author also reveals a surprising connection between Thomas Crapper and Roman Abramovich, billionaire owner of Chelsea Football Club.

Further information about Thomas Crapper's life may yet be discovered, however in light of the foregoing and as a tribute to Thomas Crapper, the Author features many previously unpublished illustrations from Thomas Crapper's 1895 catalogue and price list. This extremely rare, surviving catalogue has been in the Author's collection for many years, it was published when Thomas Crapper was at his peak, a wealthy man and content with his achievements. The catalogue is, in essence, a pictorial record of British social and industrial history, which affords an insight into the real world of Thomas Crapper.

Thomas Crapper's 1895 catalogue is bound in crimson and prominently features the gold embossed Prince of Wales feathers emblem. It is unashamedly extravagant, leaving clients and competitors alike in no doubt as to Thomas Crapper's royal credentials. A fitting tribute to The Master Plumber, Sanitary Engineer and King of the Bathroom.

Ornamental Lavatories.

No. 56

No. 57

No. 56. Superior Enamelled Cast Iron Stand, Mirror, Shelf, and Brass Towel Rail,
White Ware Basin, 26½ins. by 20½ins., with Skirting, Brass Hot and Cold
Supply Valves, and Pull-up Waste £9 7 6

„ 57. Ditto. with **Tiled Back**, Valves, and Washer and Plug ... 7 15 6

If with **Plated Fittings** ... ... Extra £0 7 6
If with **White and Gold Basin** ... „ 1 11 6

ACKNOWLEDGEMENTS

To all who have assisted or willingly volunteered information whether or not included in this book, I wish to express my sincere thanks.

Any new book about Thomas Crapper first and foremost owes a debt of gratitude to the late Wallace Reyburn who, in his book *Flushed with Pride* published in 1969, drew the modern world's attention to an almost forgotten Thomas Crapper. Reyburn's book has been reprinted many times, most recently a revised edition in 2011 which incorporates a foreword and epilogue by Simon Kirby, Chairman of Thomas Crapper & Co Ltd.

The Author particularly wishes to acknowledge the kind support and co-operation of Simon Kirby, who graciously gave up his time and unselfishly offered previously unpublished information, photographs and illustrations, many of which are included within this book.

Last, but not least, I wish to thank my wife Jacqueline for her patience and assistance.

The photographic images, illustrations and prints featured in this book are from the Author's own collection, unless otherwise stated.

Parish Church of St Nicholas, Thorne, Yorkshire
Thomas Crapper was baptised here on the 28th September 1836

CHAPTER

YORKSHIRE ROOTS

Thomas Crapper was born in 1836 at Thorne Quay a small hamlet known locally as Waterside, some ¾ mile from the ancient parish of Thorne, West Riding, Yorkshire. William White's 1837 Gazetter of Yorkshire describes Thorne as *'a small but busy market town and river port, some eleven miles North of Doncaster and thirty miles south of the Cathedral City of York.'*

Thomas Crapper was baptised at the Parish Church of St Nicholas, Thorne on the 28th September 1836, he was the seventh son born to Charles and Sarah Crapper who were parents to eight sons. In the folklore of some cultures a seventh son is born with special abilities or magic powers, for example in Ireland a gifted healer, whilst in the Spanish Americas he is believed to be a werewolf!

Thomas's father Charles Crapper was born at Hatfield, a village located near the River Don some three miles from Thorne. Charles was the eldest son of twelve children born to George Crapper, an innkeeper and his wife Ann. Charles and his siblings almost certainly attended the Travis Free Grammar School at Hatfield where the boys must have received a good education, judging by their later known occupations. This was a time when many boys in rural communities were uneducated and destined to become agricultural labourers. A variant of Crapper is Cropper – cropping is an old agricultural term, to cut and harvest a crop.

Charles Crapper had a successful career as a mariner and rose to the rank of captain. Charles's maritime commands included the Don steam packet, the John Bull steam packet as well as an aquabus sailing between Thorne and Goole. P L Scowcroft, author of *Packet Boats from Thorne* 1809-1860 referred to Captain Crapper as *'one of the great names of Thorne.'* Between 1845-1854 Charles Crapper was registered in the Merchant Navy, ticket number 103726.

Captain Charles Crapper
1796-1873

Father of Thomas Crapper

Sarah Crapper nee Green
1796-1873

Wife of Captain Charles Crapper
Mother of Thomas Crapper

Photographs courtesy of Thomas Crapper & Company Ltd.

It is possible he served in the Royal Navy, however no record has to date been discovered.

Charles's siblings were: William Crapper a mason at Brightside Bierlow, Bryan Crapper a boiler maker at Hull, John Crapper a draper and clothier at Rotherham, Benjamin Crapper a bookkeeper at York but most significant of all James Crapper born 1808 who was a plumber. James Crapper is listed in Pigot's 1834 Directory of Yorkshire as a plumber at Hatfield and in The Post Office Directory of 1837 as a plumber, glazier & painter, The Green, Thorne. It is probable that James Crapper was apprenticed to William Swallow who was listed in Baine's Directory 1822 as a plumber & glazier of Thorne. Swallow also appears in Pigot's 1829 and 1834 directories as a plumber, The Green, Thorne. Some time between 1835-36 James Crapper took over Swallow's plumbing business when, it appears, William Swallow became an employee of James Crapper. William Swallow, journeyman plumber died at Thorne in 1847. More about James Crapper later.

Historically the area around Thorne was wetland marsh until drained in 1630 by Sir Cornelius Vermuyden. The opening of the Staniforth & Keadby Canal in 1802 contributed to the town's prosperity resulting in a considerable trade in timber, iron, coal and turf to Hull and York, which together with farming and boat building provided much employment.

A fair description of Thorne where a young Thomas Crapper grew up is provided courtesy of White's 1837 Directory which lists: a Free School founded by Henry Travis in 1706, 12 Academies, 2 Banks, 4 Boat Builders, 17 Boot & Shoe Makers, 2 Breweries, 31 Farmers, 15 Inns & Taverns, 12 Beer Houses, 3 Nail Makers, 2 Rope & Flax Dressers, 3 Sail Makers, 17 Shopkeepers and 3 Plumbers. Religion was well provided for, in addition to the Parish Church of St Nicholas there were 6 Independent Chapels of a variety of denominations and 5 Sunday Schools. The population in Thorne in 1837 was c4,000.

A sailboat packet between Thorne and Hull commenced in 1809 and by 1813 it ran four times per week. Some larger vessels built at Thorne during the 1820s were used for the coasting trade also a few small warships were built for the Navy. From 1822 a wooden paddle steamship, the Kingston, built by Pearson's Shipyard at Thorne, owned by the Hull Steam Packet Company operated between Thorne and London once per week. By 1841 the schooner Fanny also launched by Pearson operated between Thorne and

Crapper Family Tree

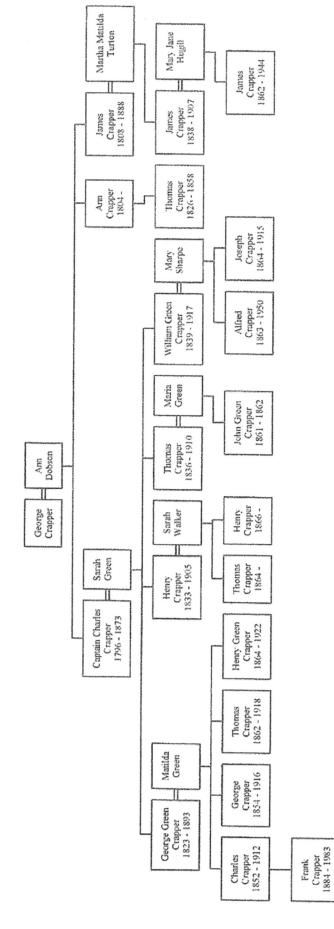

17 Crappers were engaged in the plumbing industry including 6 who worked for T Crapper & Co Sanitary Engineers

Abridged Tree

10

Stanton's Wharf, Tooley Street, London every Wednesday and Saturday, full or not.

Several packet steamboat companies operated between Thorne and Hull on a daily basis. Rivalry amongst stagecoach companies travelling between Doncaster and Hull intensified when competing against travel in the comparative comfort of the steam packet boats. Coaches raced each other for business, tragically in 1839 the John Bull coach overturned at Thorne killing one and injuring a further six passengers. Many steamboats and stagecoaches were owned and operated via local coaching inns, landowners and boat builders in Thorne. The final nail in the coffin for Thorne's steam packet boat trade came in 1856 with the arrival of South Yorkshire Railway's passenger steam trains. A few steam boats continued between Thorne and Hull until 1860 when the great days of The Don steam boat river travel came to an end.

The tiny hamlet of Waterside grew from just a few cottages in the late 18th century to 87 houses in 1851 plus 3 inns and a population of 240, although 24 houses were already uninhabited. The census of 1861 records just 57 houses of which 22 were unoccupied, the population falling dramatically during the decade. This decline reflected the end of a golden era of steam boat travel. These stark statistics help to explain why Thomas Crapper and many of his family left Yorkshire and settled in London whilst others migrated even further afield.

The 1851 census for Waterside, Thorne enumerates nine persons resident in the Crapper household including Thomas Crapper, age 14, described as a scholar, his father Charles, 56, steamboat captain, Sarah his mother, occupation: landlady, plus three brothers, two nieces and one servant. The fact Thomas Crapper is noted as 14 years of age and a scholar, is significant, when many children as young as 8 were forced to work in order to help their family survive, clearly indicates the Crapper family were in secure circumstances and even able to afford a live-in servant.

Sarah Crapper, landlady, ran a rooming or lodging house, not a public house. There were three taverns situated in Waterside: the John Bull Inn, the Rodney and the Neptune. The Crapper family occupied the cottage next to the Rodney Tavern, whose landlord for many years, was Robert Tomlinson.

The Industrial Revolution started in Great Britain during the 1770s and continued for much of the 19th century, a period when Britain dominated the

No. 58.

Ornamental Lavatories.

No. 58.

Enamelled Cast Iron Stand with Tiled Back, White Ware Basin, 27ins. by 21ins., with Silent Supply Valves and Pull-up Waste,

Complete as shewn

£6 10 6

If with Plated Fittings

Extra 7/6

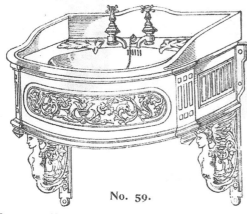

No. 59.

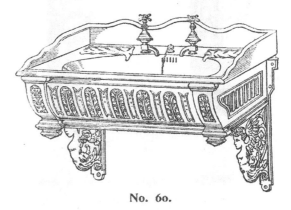

No. 60.

Nos. 59.	Enamelled Cast Iron Stand, White Ware Basin with Skirting, 16½ins. by 20½ins., Brass Hot and Cold Supply Valves, and Washer, Plug and Chain ...			...	...	...	...	£3 17 6
60.	Ditto, ditto, ditto	...	...	...	...	...	...	3 17 6
	If with Plated Fittings	...	...	...	Extra	£0 7 6		
	If with White and Gold Basin	...	...	...	,,	1 9 6		

world economy. A financial crisis in 1837 resulted in a hard recession which lasted for several years causing hardship and unemployment for many citizens, particularly in the north of England.

Thomas Crapper's uncle, James Crapper, advertised his business for sale in The Leeds Mercury on the 6th February 1841; *'TO PAINTERS - TO BE DISPOSED OF, with immediate possession, an old established BUSINESS, which has been successfully carried on for Thirty Years. In a Market Town in the West Riding of the County of York. Valuation of Stock and Fixtures is but small. A Front and good Workshop, and a comfortable Dwelling House are attached, with a good Garden etc. Apply personally, or by letter, (pre-paid) to Mr James Crapper. Plumber, Glazier and Painter, Thorne. Feb 1841.'*

James Crapper sold his Thorne plumbing business in 1841 and with his wife Martha and young son James junior age 2 migrated to Canada, which at that time was a popular destination for many British emigrants. The family all survived the long, arduous and dangerous voyage and upon arrival settled in Toronto, where it is believed James Crapper was the first plumber to successfully set up in business.

Coincidentally, prior to 1836, Toronto was in fact known as York and featured a great wetland marsh fed by the River Don! James Crapper must have felt 'right at home'. The Canadian Crappers thrived, James lived to be 80 years of age and his son James junior (1838-1907) also a plumber handed the business on to James Crapper III (1864-1944). There is still a James Crapper Plumbing & Heating Sales in Toronto, Canada in 2014.

The Crapper family plumbing connection with Thorne did not end in 1841 with the departure of James Crapper. Tom Crapper born 1826 at Hatfield, the illegitimate son of Ann Crapper, a nephew of James and cousin of George and Thomas, initially worked for his uncle John Crapper, clothier, however he soon turned to plumbing, believed to have been apprenticed to George Pattrick, plumber and glazier of Finkle Street and Orchard Street, Thorne. Tom Crapper married Ann Caroline Gunnee in 1850, she was the daughter of George Gunnee, painter and glazier, King Street, Thorne. She was also a niece of George Pattrick who died in 1847.

On Saturday 11th January 1851 a report appeared in the Sheffield Independent Newspaper, it stated *'Thomas alias TOM CRAPPER, 26 was charged with breaking into the shop of George Gunnee, at Thorne, and stealing therefrom 40lbs of lead piping, sentence TWO MONTHS.'* It appears Tom was

Pedestal Wash=down Closet.

"The Rapidus."

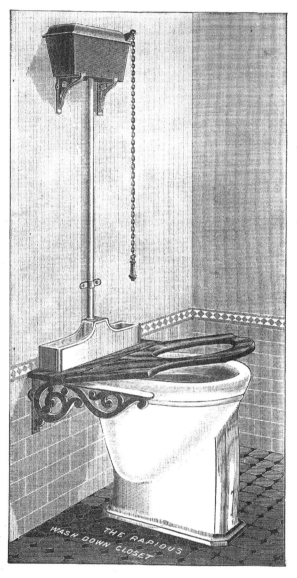

An

Economic

and

Effective

Wash-down

Closet.

Combining

W.C.,

Urinal

and

Slop Sink.

No. 166.

No. 166. White or Ivory Basin and Trap in one piece with Paper Box, Mahogany
or Walnut Seat, and 3-gallon Syphon Water Waste Preventer with
China Pull (Supply Pipe not included) £3 17 6

Basin and Paper Box Printed outside   Extra 9/6

the black sheep in the family! Tom Crapper age 25 is enumerated in the 1851 census as a plumber and glazier, Finkle Street, Thorne. Tom died in 1858 aged 32 leaving three infant sons; George, Tom and Samuel fatherless. His namesake (unrelated) Tom Crapper 18 was less fortunate when convicted of stealing a pair of shoes for which he was sentenced to 7 years transportation. (Leeds Times 9th January 1847)

Whilst he was growing up Thomas Crapper's playground was the river bank, a natural attraction for any young lad, no doubt he participated in local activities like fishing, boating and swimming. Thorne Waterside accommodated the local working community who were predominantly involved in river transportation, boat building, inns, shops, lodging houses etc. Young Thomas Crapper was surrounded by the sight, sound and smells of manufacturing, mechanical machinery, saw mills, the blacksmith's anvil and the steamboats' chimneys belching out clouds of black acrid smoke. Thomas would also have witnessed first hand the excitement of a new boat launch when thousands attended from miles around to join in a rare day of celebration and entertainment.

Thomas must have been fascinated when observing the plumbers' skills whilst they installed the complex steam boilers and equipment at the boatyards. Who knows, he probably watched a Downtown's ship's closet being installed which fired up his life-long interest in sanitation.

By the time Thomas Crapper was a teenager his uncle, two of his older brothers and his cousin were already working plumbers, no doubt influencing Thomas's decision to take up plumbing as his career, a wise choice, as it was an occupation in which he was destined to become 'The Master'.

Judging by Thomas Crapper's range of sanitation related patents and designs he clearly had a very inventive mind and a determination to manufacture improved sanitary appliances to the benefit of everyone. The Yorkshire lad from Thorne Quay certainly made an important contribution and a lasting legacy which survives to this day.

Of Thomas Crapper's seven brothers, George, Henry and William became plumbers in London, they each followed in their Uncle James Crapper's chosen trade, although not all of them were successful. Brothers John, Robert and James followed their father's occupation becoming mariners. Robert rose to become a ship's captain like his father, whilst John made first mate. Charles Crapper junior followed his Uncle John Crapper's trade and became

George Green Crapper
1823-1893

Older Brother of Thomas Crapper

Matilda Crapper nee Green
1832-1894

Wife of George Crapper

Photographs courtesy of Thomas Crapper & Company Ltd.

a tailor initially in Halifax, Yorkshire, however by 1860 he too had moved to London.

George Crapper (1823-1893) is believed to have been an apprentice plumber in 1837 to his uncle James Crapper, however when James migrated to Canada in 1841 he transferred to another Master Plumber situated at Hull in order to complete his apprenticeship. The 1841 census enumerates George Crapper, apprentice plumber lodging in the home of George Shirtliffe, a ship's pilot working on the busy Humber who would almost certainly have been acquainted with Captain Charles Crapper.

George Crapper, upon completion of his apprenticeship, decided to seek his fortune in the capital and relocated to London c1843, a time when plumbers were in great demand. In 1851 he was residing at the home of Francis Griffiths, gentleman, 2 John's Place, Marlborough Road, Chelsea, age 28, occupation: plumber.

A notice appeared in the Hull Packet & East Riding Times on 25th April 1851 *'Marriage at the Holy Trinity Church, Hull by the Rev W B Hopkins MA 13th April 1851 at this place. Mr G Crapper of London eldest son of Captain Crapper of Thorne to Miss Matilda Green, late stewardess of the Don Steam Boat plying between Thorne and Hull.'* Matilda Green was born at Scratby, Norfolk in 1832, she was a niece of George Crapper's mother Sarah Crapper nee Green. In the 1851 census Matilda is lodging with the Crapper family at Thorne.

George and Matilda made their home in Chelsea, London, soon other family members followed them. George rapidly established a successful plumbing business necessitating an increase in his workforce to meet demand, creating a perfect opportunity for young Thomas Crapper to follow in his brother's and his uncle James Crapper's footsteps. The Thorne plumbing business owned by James Crapper was established in 1811, over 200 years ago. Exactly fifty years later Thomas Crapper established his own successful plumbing business in Chelsea, London.

Tucked away in a quiet backwater, a short distance from a busy motorway, lies the village of Waterside near Thorne. Landmarks familiar to a young Thomas Crapper 170 years ago are still recognisable today. The John Bull Inn remains a welcoming focal point where a photograph of Thomas Crapper, a local boy made good, proudly adorns the wall. Some of Waterside's 19th century two and three storey cottages still exist, although by the 1880s the village had become largely depopulated almost to the point of near desolation.

Thorne Quay, Waterside early 1900s

Thorne Quay, Waterside c1940s

Photographs courtesy of the John Bull Inn

The old stone quay and a restructured brick built warehouse survive, still visible are the steps leading down to where several steam packet boats once moored. The River Don was diverted many years ago leaving Waterside high and dry, the river bed and bank, now overgrown, provide a peaceful sanctuary for birds and wildlife. It is easy to picture Waterside in bygone days as a bustling river port with stage coach and steam boat passengers, cargo transportation and a thriving boat building industry surrounded by the tranquility of unspoilt farmland.

Thomas Crapper is not the only famous past resident of Waterside, Lesley Garrett, soprano, also grew up in one of the old quayside cottages that once overlooked the great River Don.

The River Thames, London 1841
Stanton's Wharf is situated bottom right.

CHAPTER

LONDON APPRENTICE

We can be fairly sure that young Thomas Crapper travelled from his birthplace Thorne Quay aboard the schooner Fanny directly to Stanton's Wharf, Tooley Street, Southwark, London c1852. The sea journey took between 24-36 hours. His father, Captain Charles Crapper, no doubt arranged for local Thorne crew members to keep an eye on his young son. It is not difficult to imagine young Thomas Crapper's excitement whilst he stood on the ship's deck as it sailed up the bustling River Thames to its destination only a mile or so from Chelsea and his brother George.

Thomas came from an environment where he knew everyone and everyone knew him, he was surrounded by open countryside, fresh air and Yorkshire folk. George Crapper, his older brother, surely explained to Thomas that London contained great riches but also much poverty, both extremes would be an eye opener to a young Yorkshire lad from Thorne.

The following extract is taken from a report into the problems of urban expansion in England 1852;

'Fifty to sixty houses, no water. All dirty, pallid, diseased and some idiots. All so bad as to be indescribable; a man almost dying; a woman with half a face; children almost devoured with filth; prostitutes and thieves; Privies almost invariably against the houses and people complain of the stench. The physical and moral condition of this place is indescribable.'

London had districts of great deprivation, the like of which Thomas Crapper had never imagined. Soon he would play his part in helping to improve the sanitary conditions and the quality of life for many thousands of people, rich and poor alike.

After George and Matilda married they made their home at 6 Keppel Terrace, close to Marlborough Road, Chelsea. Their first born Charles George was baptised at St Luke's Church on 7th March 1852. George and Matilda had nine children, six boys and three girls. George was doing well and soon

The London Apprentice
Made of lead in 1769, the figure stood in Ye-Olde Plumbers Shop
of Dent & Hellyer, Newcastle Street, London, destroyed in a fire in 1903.
Figure incorporated within the trade mark of Beard, Dent & Hellyer 1874.

The Apprentice Toff
Lead casting of Eton School Boy c1860s.
Cast by John Bolding & Sons, London.

became a respected Master Plumber employing eight men. As his family increased he moved house in 1854 to 9 Little Orford Street and again in 1858 to 62 Walton Street where the family remained until 1869, the rent was 50/- (fifty shillings) per year.

Thomas knew Matilda, she was a niece of his mother and lived with them at their Thorne home whilst working for Captain Crapper as a stewardess aboard the Don steam packet boat. For the next eight years Thomas resided with George and Matilda in Chelsea, London.

Thomas Crapper was an apprentice plumber to his brother George for five years, 1852-57. It was common practice during the 19th century for a relative, usually their father, older brother or uncle, to take on a younger male relation often without any formal indenture, this saved the payment of stamp duty. Thomas was very fortunate, an apprenticeship in plumbing was highly valued. A skilled craftsman who was taught a specialised craft or 'mystery' via an apprenticeship was regarded far higher than unskilled labourers, reliant on muscle power for a living.

Another route to an apprenticeship included payment by the boy's parent or guardian to the Master in order that he takes him on. The final route to an apprenticeship was for pauper children in the care of the parish workhouse, with the parish funding the Master's fees. Apprenticeships were not only reserved for the crafts, attorneys, surgeons, apothecaries etc were often trained via an apprenticeship too.

The Author has, in his collection, a typical indenture dated 1840 for apprentice James Alfred Kettle bound to Edward Haggar, Master Plumber, Glazier and Painter of Ipswich, for a term of 4 years 5 months. The boy's father Joseph Kettle, a farmer, paid Edward Haggar two sums comprising £10 at the commencement of the apprenticeship and a further £10 at the start of the third year. The apprentice was to be paid 1/- per week during the 5 month probationary period followed by 2/- per week year 1, 3/- per week year 2, 4/- per week year 3 and 6/- per week year 4.

'The apprentice shall learn his Master's art faithfully, shall serve his Master, his secrets keep, his lawful commands everywhere gladly do. The goods of his Master he shall not waste, Shall not haunt Taverns, Inns or Play houses. Shall not commit fornication nor contract matrimony within the said term. Shall not play cards or dice tables or other unlawful games. Shall not buy or sell, nor absent himself from his Master's service day or night

Lead Junctions.

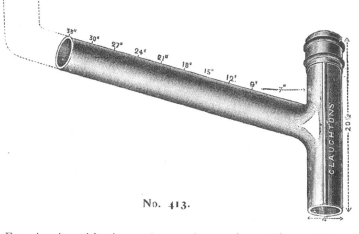

No. 413.

No. 414.

For 4 in. pipe, with 7 in.   9 in.   12 in.   15 in.   18 in.   21 in.   24 in.   27 in.   30 in.   33 in.   36 in. Arm.

No. 413.   ...   7/9   8/7   9/3   9/11   10/7   11/3   12/1   12/9   13/5   14/1   14/9

,, 414.   For 4 in. pipe, with two 7 in. Arms   ...   10/9 ; with 9 in. Arms   ...   12/6

Either Arm can be extended   Extra 8d. every 3 in.

Patent Lead and Iron Connections.

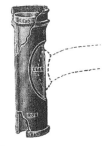

No. 415.

No. 415.   2½ in. for Branching in either 1½ or 2 in. Lead   4/6

4 in.   ,,   ,,   3½ or 4 in. ,,   6/9

W.C. and Soil Pipe Junction.

No. 415a.

Patent Lead to Iron Junction, with Lead
Arm up to 24 in. in length, and Screwed
Lead Connection for Trap of W.C. 20/9

Drawn Lead Pedestal Closet Traps.

No. 416.

P Trap, equal to 6 lb. Lead

            3½ in.   4 in.

            9/9   11/-

Ditto, equal to 8 lb.

   Lead   ...   11/3   12/6

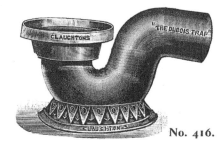

No. 416.

No. 416a.

S Trap, equal to 6 lb. Lead

            3½ in.   4 in.

            11/6   12/9

Ditto, equal to 8 lb.

   Lead   ...   13/-   14/3

unlawfully. But in all things as a faithful apprentice he shall behave himself towards his Master. The Master shall provide the apprentice with sufficient meat, drink and lodgings and all other necessaries. The Master shall teach and instruct unto the apprentice which he useth by best means during the said term.' The indenture is signed and dated by all three parties.

Work undertaken by Plumbers 1827-1901:

Lead water services	Steam installations
Lead gas services / lighting	Hot water heating
Lead cisterns	Below ground drainage
Lead baths and sinks	Speaking tubes communications
Lead water closet pans and basins	Electric bell systems
Lead soil and waste pipework	Alarm systems
Lead syphons and traps	Pneumatic bells
Lead roofing, guttering	Ventilation
Lead pumps	Fireplace installations
Lead coffins	Painting and decorating
Lead decorative work	Glazing
Copper hot and cold water services	Sign writing
Plumbing on ships	Coppersmithing
Plumbing on trains	Boilers and cylinders

Specialist trades that have evolved from plumbers' work include: lead worker, glazier, painter & decorator, gas engineer, boiler engineer, air conditioning engineer, refrigeration engineer, heating engineer, telephone engineer, electrician, alarm engineer, sign writer, solar heating engineer, bathroom and kitchen specialist.

Lavatory Basins.

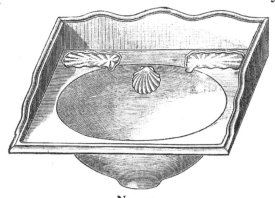

No. 79.

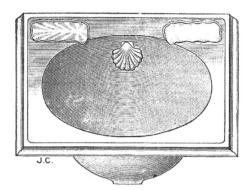

No. 80.

No. 79. Oblong, with Back and Sides, 26½ × 20½ins.   Oval Basin   21 × 15ins. £1 14 6
      27 × 19 ,,   ,,   20½ × 14 ,, ... 1 14 6
      25 × 18 ,,   ,,   19 × 13 ,, ... 1 10 6
      22 × 16 ,,   ,,   17 × 11 ,, ... 1 4 6
,, 80. Oblong ...   ...   26½ × 20½ ,,   ,,   21 × 15 ,, ... 1 10 0
      25 × 18 ,,   ,,   19 × 13 ,, ... 1 6 6
      22 × 17 ,,   ,,   17 × 12 ,, ... 1 1 6

Also made with Tap Holes and Grid Overflow.

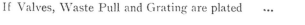

No. 81.

No. 81. Ornamental Lavatory with Skirting, 27 × 22ins., with ½-in. Brass Standard Valves and Improved
   Standing Waste £4 7 6
     If Valves, Waste Pull and Grating are plated   ...   ...   Extra 8/6

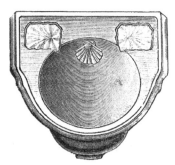

No. 82.

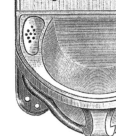

No. 84.

No. 82. Round Front, 18 × 17ins., Round Basin, 12ins. 16/3
,, 83. Ditto, with Back 20/-
,, 84. Semi-Plug Basin, with 3in. Back, Patent Overflow and Drained Soaps, 18½ × 12ins.,
     Basin, 11½ × 10ins. 17/6

During the period 1843-1860, after George and Thomas Crapper arrived in London, the following notable events occurred:

1848　　　　The Public Health Act, established national control of plumbing and sanitation.

1848-50　　25,000 deaths occurred in London from cholera. Dr John Snow, a physician, provided evidence to show that cholera deaths were caused by people drinking water contaminated by sewage.

1851　　　　London hosted The Great Exhibition at Crystal Palace. The nation showcased industry, inventions and manufacturing. George Jennings provided public conveniences to the exhibition for which he was allowed to charge 1d (one penny) to use the toilet.

1858　　　　The summer of 'The Great Stink' – the River Thames was polluted with sewage, Parliament was forced to expedite construction of an adequate system of sewage disposal for London.

Thomas Crapper left Yorkshire in the hope of making his fortune, he could not have chosen a better place to set up in business than Chelsea, London. The borough was home to many influential, wealthy people, it contained numerous mansions, high quality residential and commercial properties. Chelsea is located in close proximity to central London which was a source of continuous work, new construction and refurbishment. Chelsea was an affluent area which was well governed, containing relatively few slum properties and in consequence low numbers of poor residents, unlike many other areas of London. Consequently, Thomas Crapper gained a substantial range of plumbing experience during his apprenticeship.

George Crapper progressed from local plumber to property developer, he became a man of substance, influence and a respected member of the Chelsea community. The Chelsea Vestry records of 1858-59 confirm George Crapper was the appointed plumber and glazing contractor for the area. In 1863 George Crapper, 62 Walton Street, was elected as a vestryman representing Hans Town Ward. Vestryman George Crapper was a member

THOMAS CRAPPER & Co.,

Manufacturers and Patentees of
Sanitary Appliances.

MARLBOROUGH WORKS,

Marlborough Road, Chelsea, London

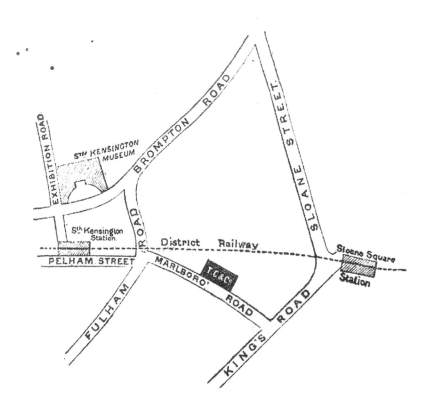

Five minutes walk from either South Kensington or Sloane Square Station, District Railway.

Image from Thomas Crapper & Co Catalogue 1892-93

Courtesy of Thomas Crapper & Company Ltd.

of the Committee of Works. In April 1864 a meeting was held to discuss the proposed railway route through Chelsea which would affect Marlborough Road resulting in all the old established tradesmen being displaced. George Crapper voted against the proposal which was defeated, 28 against and 15 for.

George Crapper moved away from Chelsea in 1869 to his newly constructed Thornsett House, Heathfield Road, Mill Hill Park, Acton, later described as *'a handsome modern detached, double fronted residence with large well stocked gardens, 12 good rooms, bath and offices and every convenience.'*

George Crapper had a major influence on Thomas Crapper's future career. It was George who encouraged and helped Thomas establish his business, later they operated together as property developers. George became a Master Builder, his two sons Charles and Thomas inherited his building, plumbing & decorating business.

All available evidence indicates that George Crapper was indeed the '& Co' in Thomas Crapper & Co.

Thomas Crapper
1836-1910

Maria Crapper nee Green
1836-1902

Wife of Thomas Crapper

Photographs courtesy of Thomas Crapper & Company Ltd.

3

CHAPTER

CHELSEA PLUMBER

Upon completion of his apprenticeship c1857 Thomas continued working for his brother George as a journeyman plumber, gaining further valuable experience of installation and repair work throughout Chelsea and Kensington. Like other members of his family, Thomas held aspirations of establishing his own plumbing business.

On 26th July 1860 at Trinity Church, Upper Chelsea, Thomas Crapper married Maria Green, his cousin, they were both 23 years of age. Maria was the daughter of Robert Green, a farmer at Hemsby, Norfolk. Robert was the brother of Sarah Crapper, Thomas's mother. Thomas and Maria made their first home together at 3 Marlborough Cottages, a Georgian property within a small row of five cottages in College Street, Chelsea. The house was located in a good residential area where previous occupants were John Josolyne, commission agent and his wife Eliza, a schoolmistress for young ladies. Other former tenants of Marlborough Cottages included William Wheally, a pianoforte maker and William Harvey, an artist in water colours.

Thomas Crapper's parents arrived from Yorkshire in the summer of 1860, they joined the exodus to pastures new when the Thorne steamboat industry diminished. Captain Charles Crapper was subsequently described as an annuitant, clearly a reward for his long service in the Merchant Navy. George Crapper accommodated his parents and brother Charles junior at 62 Walton Street, Chelsea. At this time George and Matilda had three children and Matilda was heavily pregnant with a fourth.

Sadly Robert Green Crapper, the child George and Matilda welcomed into the world, died January 1861 and was interred at Brompton Cemetery 2nd February age just 5 months. George and Matilda had nine children in total, their sons George and Henry worked for Thomas Crapper & Co. Interestingly also recorded in Brompton's burials' register on 2nd February was one Charles Dickins, age 44!

Wash=down W.C. Basins and Traps.

"The Ovington."

No. 198. White Ware Basin,

and Lead P Trap £1 1 0

If with Printed Basin Extra 2/-

,, S Trap ... ,, 1/9

No. 198.

With Lead Trap.

"The Chelsea"
With County Council Trap.

 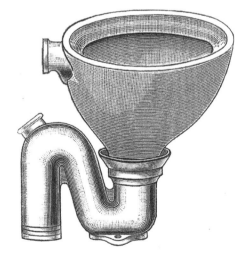

No. 199.

No. 200.

No. 199.	Cane and White	...	5/9		No. 200.	Cane and White	...	7/3
	White	...	...	7/9		White	...	9/3
	Ditto and Printed	...	9/3			Ditto and Printed	...	10/9

The next mention of Thomas and Maria Crapper is in the 1861 census for Chelsea, whilst they remain resident at 3 Marlborough Cottages, they have an addition to their family, a son John Green Crapper age one month. Thomas and his family moved to nearby 42 Robert Street, Chelsea, a property large enough for him to establish himself in business, enabling him to realise his ambition.

John Green Crapper was baptised on the 23rd June 1861 at St Luke's Church, Chelsea. At about the same time Thomas Crapper's younger brother, William Green Crapper arrived in Chelsea from Yorkshire with his wife Mary and baby Robert William Crapper. Thomas accommodated William and his family at his house in Robert Street until they found accommodation of their own. It is thought William was Thomas Crapper's first employee. Four Crapper brothers relocated to Chelsea plus their parents, one other brother was yet to venture southwards.

Thomas Crapper, plumber and glazier, was awarded his first contract by the Chelsea Board of Guardians on the 19th September 1861, the contract was for six months. On the 20th March 1862 Mr T Crapper of Robert Street was successful in his tender to the Chelsea Vestry and was awarded a six month contract for plumbing and glazing.

Tragically William and Mary Crapper's son Robert William Crapper was found dead at 42 Robert Street on the 19th March 1862, age 9 months, a Coroner's Inquest held 21st March recorded the cause of death as *'congestion of blood and effusion of the serum in the head'* (a brain hemorrhage). Robert William Crapper was buried on the 22nd March at Brompton Cemetery.

A few weeks later, on 11th April, Thomas and Maria had the devastating and heartbreaking experience of burying their own precious son John Green Crapper who died 8th April 1862 age 1 year 1 month, he was also interred at Brompton Cemetery. His cause of death was recorded as *'mesenteric gland atrophy'*. The death certificate states Thomas Crapper, Plumber (Master), 42 Robert Street, Chelsea as the father and informant. Thomas and Maria had no further children.

William and Mary moved out of Robert Street into 29 Hasker Street, Chelsea, where on 7th February 1863 their son Alfred William Crapper was born. William and Mary were parents to nine children, at least two of whom, Alfred and Joseph, worked for Thomas Crapper & Co.

During the period 1860-1866 Thomas Crapper experienced the highs of personal happiness and the devastation of family tragedy, throughout he

Valveless Water Waste Preventers.

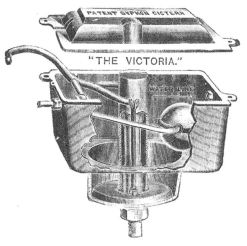

"THE VICTORIA."

No. 226.

Noiseless.

No Valves or Rubber.

No Drip or Splashing.

Right or Left Hand.

Approved by the

leading

Water Companies

in London and the

Provinces.

	2-gallon.	3-gallon.
No. 226. Stamped by the New River Company	17/9	
To meet the requirements of the Croydon Corporation	16/6	19/6
To suit other Metropolitan and Provincial Water Companies	14/6	18/-
Galvanized Extra 2-gallon 4/6; 3-gallon 6/-		

"The Leverett."

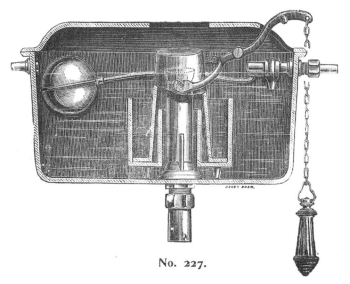

No. 227.

No. 227. Cast Iron 2-gallon, with Pull	15/9
Ditto 3-gallon, ditto	18/6

remained industrious, determined to succeed, building a growing reputation, whilst establishing himself firmly in business.

Somewhat surprisingly all three of the Crapper brothers' young sons who died, Robert Green, Robert William and John Green, were each buried in separate common graves recorded in the burial register as 12 feet deep, each were followed by 3 or 4 other unrelated persons. A few years later George Crapper purchased a private plot at Brompton Cemetery, this grave contains the mortal remains of Walter Crapper buried 17th September 1867 age 1 year, son of George and Matilda Crapper, Captain Charles Crapper buried 1st October 1873 age 77, Sarah Crapper buried 29th October 1873 age 77, George Green Crapper buried 19th January 1893 age 69 and Matilda Crapper buried 20th February 1894 age 62.

Typhoid caused the death in 1861 of Queen Victoria's husband Prince Albert along with 25,000 other London residents. Infectious diseases prevalent in England at that time were cholera, consumption (now known as tuberculosis), smallpox, measles, whooping cough and diarrhoea.

A report in The West Middlesex Advertiser regarding a meeting held on 23rd June 1863 at The Chelsea Vestry (the local authority) included details of recorded births and deaths within the parish during the previous 8 week period. Births totalled 84 of which 48 were boys and 36 girls. There were 59 deaths comprising 27 under 5 years of age of which 16 were infants under 1 year old. Deaths from measles continued to decline, they numbered 17 in the fortnight ending 4th May, 10 on the 30th May.

Thomas Crapper advertised in Simpson's 1863 Chelsea Directory, his entry read; 'T Crapper, Brass Finisher, 42 Robert Street, Chelsea.' Clear evidence Thomas was diversifying his business strategy. In the same directory George Crapper advertised as 'Plumber, 62 Walton Street, Chelsea.' Almost certainly Thomas and George Crapper pooled resources, helping each other as and when required.

The West Middlesex Advertiser, Saturday 27th June 1863, reported the minutes of the Chelsea Vestry meeting including 'Plumber – the following tenders were received; Messrs Parsons at schedule prices; Mr John Todd ¼% below schedule prices; Mr Thomas Crapper, Robert Street 17s 6d below schedule prices. The various names were put from the chair when there appeared for Mr Crapper 18, Messrs Parsons 7, Mr Todd 0. The tender of Mr Crapper is accepted.'

Cistern Pulls.

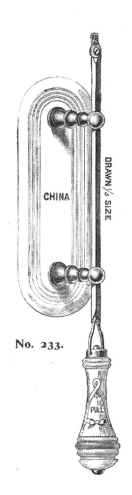

No. 233.

No. 234.

No. 235.

No. 236.

No. 233.   Cream, Blue or Black China, with Handle to match 5/9

Ditto, with Bevelled Edge Plate Glass Back with Cut Star Centre   ...   7/9

No. 234.   Crown Derby Design   2/9

No. 235.   Solid Mahogany, Walnut or Oak Block, with Moulded Edge   ...   5/3

No. 236.   Cream and Gold   ...   1/9

Twisted Flutes, Cream and Gold   ...   1/9

Twisted Flutes, Blue and Gold   ...   1/9

These important Vestry contracts paved the way and opened many doors for Thomas Crapper, leading to numerous valuable contracts for sanitary and water installations to municipal buildings, schools, barracks, mansions and eventually royalty. Thomas expanded his business taking on more plumbers and family members.

Thomas Crapper was a staunch advocate of training his own first-class plumbers via the traditional method of time served apprenticeship. It may be possible to teach someone to replace a defective part in a short space of time, however only substantial experience gained over many years can equip a qualified tradesman with the necessary skills to diagnose potential dangerous and complex faults.

Many of Crapper's employees remained with his company all their working life, whilst others e.g. Robert Marrion and Frederick Humpherson used their Crapper training experience to advance their careers. Robert Marrion was apprenticed to Thomas Crapper c1863, he was almost certainly Thomas's first apprentice plumber. As Crapper's business grew so Robert Marrion progressed to become foreman and subsequently the plumbing manager of T Crapper & Co. In 1886 Marrion left Crapper's employ and emigrated to Vancouver, British Columbia joining two of his brothers already in Canada. Robert Marrion was appointed Public Health & Plumbing Inspector for Vancouver in 1896, a position he held until 1912. He remained Vancouver's Plumbing Inspector from 1912 until his death in 1922, age 73 years.

During the early 1860s Thomas Crapper concentrated on establishing his plumbing installation business and putting food on the table, so to speak. In 1866 he committed himself financially to his newly acquired Marlboro' Works brass foundry. Around 1870 Thomas diversified further by developing his plumbing merchant business, initially supplying local plumbers, however he was soon supplying contracts all over London. In 1872 Thomas created his first bathroom showroom at numbers 50 and 52 Marlborough Road including the installation of large plate glass display windows, finally incorporating number 54 in 1885 following his acquisition of the last property in the old Buckingham Place (Marlboro' Works).

Having firmly established his, by now, extensive and profitable business Thomas, with over 25 years experience in the industry, gradually delegated responsibilities, turning his creativeness to designs and patents. Thomas Crapper did not look to the bathroom immediately instead his own practical

Cast Iron Bath with Fittings.

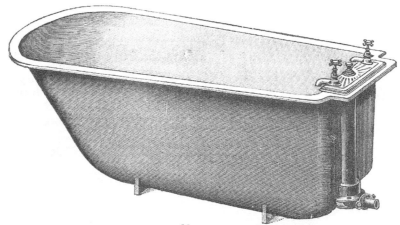

No. 11.

			5ft. 6in.			6ft. 0in.		
No. 11. Cast Iron Bath, Japanned inside and plain painted outside, with China Tray, Hot and Cold Supply Valves and Lift-up Waste, complete			£3	5	6	£3	19	6
,, 12. Ditto ditto Enamelled inside			4	8	0	5	2	6

If with Plated Fittings   ..   Extra £0 9 6

If Decorated outside and fitted with Polished Mahogany
     or Walnut Capping   ...   Extra £2 5 0

Independent Bath.

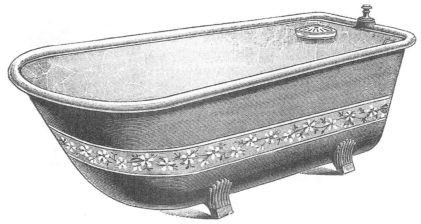

No. 13.

			5ft. 6ins.			6ft. 0in.		
No. 13. Cast Iron Bath, Japanned inside and Decorated outside, China Soap Cup and Lift-up Waste only ...			£3	11	6	£4	6	6
,, 14. Ditto ditto Enamelled inside			4	14	0	5	19	6

If Plated Pull   ...   .. Extra 1/9

experience caused him to focus his attention on improving the perilous state of the nation's underground foul drainage systems.

Thomas Crapper's first registered design, Crapper's Disconnecting Trap (1881) patented in 1888 was followed in 1894 with the Improved Kenon Disconnecting Trap. Crapper's disconnecting trap was effectively the best on the market and stocked by many leading merchants e.g. Young & Marten, The Farringdon Works Ltd / H Pontifex & Sons Ltd. Between 1881-1903 Thomas Crapper was credited with seventeen patents and designs, including novel inventions together with design improvements to inspection chambers, grease traps, ventilation, syphon cisterns, closet pans, bath traps, joints and even stair rods.

As a leading sanitary engineer Thomas Crapper's progressive contribution was cumulative and substantial, yet generally underestimated by many historians. Thomas Crapper was born too late to invent the water closet, however no sanitarian during the important 1880-1900 boom period did more than Thomas Crapper to promote improvements to the nation's sanitation and bathroom facilities, as such the Author believes Thomas Crapper is entitled to be known as the King of the Bathroom.

George Crapper 1854-1916

Robert Marr Wharam 1853-1942

Photographs courtesy of Thomas Crapper & Company Ltd.

Owners, Managers and Directors of Thomas Crapper & Co 1861-1963

Name	Duration	Title
Thomas Crapper (1836-1910)	1861-1904	Founder and owner of T Crapper & Co
Robert Marrion (1849-1922)	1863-1886	Worked for T Crapper & Co Apprentice, Foreman, Plumbing Manager
George Crapper (1854-1916)	1869-1904 1869-1885 1885-1895 1895-1904 1904-1916	Worked for T Crapper & Co Brass Finisher Brass Foundry Manager Manager of Sanitary Works Director, T Crapper & Co Ltd
Robert Marr Wharam (1853-1942)	1870-1904 1870-1887 1887-1904 1904-1942	Worked for T Crapper & Co Commercial Clerk Junior Partner / Manager Managing Director, T Crapper & Co Ltd
Henry Crapper (1864-1922)	1878-1904 1878-1891 1891-1904 1904-1922	Worked for T Crapper & Co Brass Worker Foreman Plumber and Brass Finisher Foundry Foreman and Brass Finisher T Crapper & Co Ltd
William Gubbins (1863-1929)	1878-1904 1904-1929	Commercial Clerk, T Crapper & Co Company Secretary, T Crapper & Co Ltd
Robert Gillingham Wharam (1885-1967)	1904-1942 1942-1963	Director, T Crapper & Co Ltd Chairman, T Crapper & Co Ltd
Robson Barrett (1889-1977)	1904-1963	Worked for T Crapper & Co Ltd Salesman, Clerk, Manager, M D

Gun Metal Cocks.

Gland Bib Cocks

(S.E. and Ebony Lever)

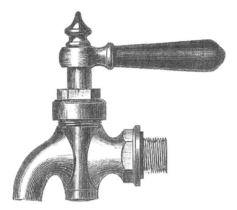

No. 306.

	$\frac{1}{2}$ in.	$\frac{3}{4}$ in.	1 in.
No. 306. ...	5/9	8/6	12/6

Shield Pattern Bib Cocks

(S.B., S.E. and Ebony Lever)

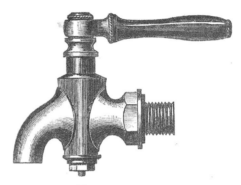

No. 307.

	$\frac{1}{2}$ in.	$\frac{3}{4}$ in.	1 in.
No. 307. ...	5/9	8/6	12/6

Screw=down Bib Cocks

With Ebony Wheel Top.

No. 308.

Shield Pattern Bib Cocks

No. 309.

			$\frac{3}{8}$ in.	$\frac{1}{2}$ in.	$\frac{3}{4}$ in.	1 in.
No. 308.	Screwed for Iron ...	...	3/4	3/9	4/11	8/11
	Screw Boss ...	...	3/7	4/1	5/5	9/8
,, 309.	Screwed for Iron ...	...	4/2	4/9	6/9	10/6

Thomas Crapper & Co Ltd – List of Shareholders 1904

Robert Marr Wharam Managing Director	3,667 shares (51%)
George Crapper Director	1,881 shares (25%)
Robert Gillingham Wharam Director	1,701 shares (23%)
Thomas Crapper Former Owner	1 nominal share
William Gubbins Company Secretary	1 nominal share
Richard Woodgate Commercial Clerk	1 nominal share
Alfred Spice Architectural Draughtsman	1 nominal share
George Richard Wharam Draper's Assistant	1 nominal share

➤ William Gubbins commenced work for Thomas Crapper & Co c1878 as a commercial clerk. He lived at 27 Green Street, Chelsea, later at 4 Geraldine Road, Wandsworth, Battersea. He died in 1929.

➤ Richard Woodgate resided next door to George Crapper at 21 Gorst Road, Wandsworth, Battersea.

➤ Alfred Spice resided at 86 Bridge Road West, Battersea.

➤ George Richard Wharam, born 1852 at Thorne, brother of R M Wharam, bachelor, resided at 141 Warwick Road, Kensington with his spinster sister Annie Evangeline Wharam. He died in 1936.

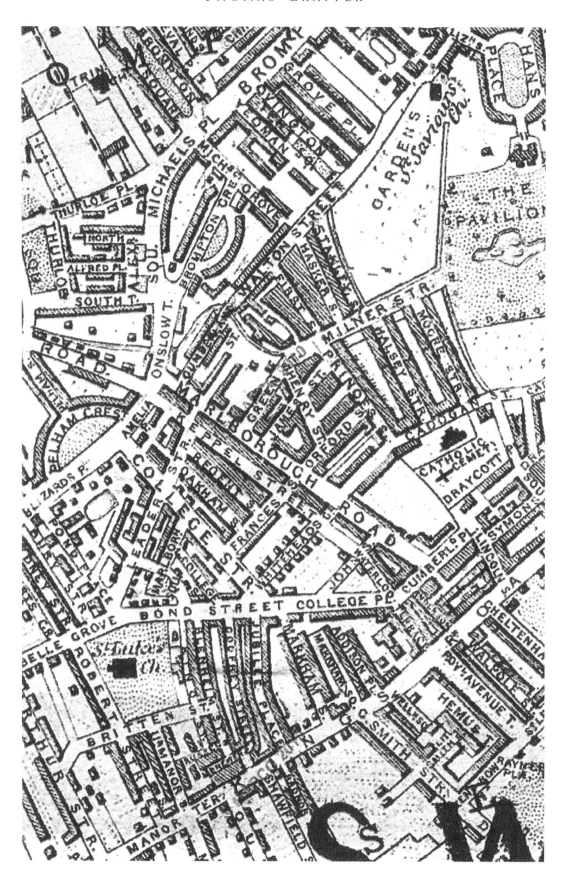

Map of Chelsea 1857

Thomas Crapper's Chelsea Residential Addresses

Address	Year	Status
Keppel Terrace	1853-1854	Living with brother George.
9 Little Orford Street	1854-1858	" " " "
62 Walton Place	1858	" " " "
62 Walton Street	1858-1860	" " " "
3 Marlborough Cottages, College Street	1860-1861	First married address.
42 Robert Street	1861-1867	Home and business.

Thomas Crapper Residential & Investment Properties Outside Chelsea

Address	Year	Status
1 Middleton Road, Battersea	1867	Purchased May 1867 for £174.
8 & 10 St Mark's Villas, Middleton Road	1873	Developed by Thomas who moved into No.8 in 1874.
21 Powis Square, Brighton	1890-1895	Thomas commuted to London.
12 Thornsett Road, Anerley, Penge, Kent	1895-1910	This property remained in the Crapper family until 1944.
Middleton Road, Battersea	1870s	Thomas & George developed other properties in this location.
51, 53, 61 & 63 The Gwynne Estate, Battersea	1881	Thomas Crapper leased these plots for investment/ development.
Rear of 12 Thornsett Road, Anerley, Penge, Kent	1895-1910	Thomas acquired a building plot at the rear of his house which he used as a vegetable garden, sold after his death in 1910.

Ball Valves with Copper Balls.

Croydon Valves

(Stamped by New River Company).

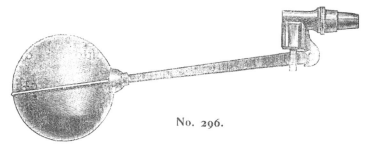

No. 296.

	$\frac{1}{2}$ in.	$\frac{3}{4}$ in.	1 in.	$1\frac{1}{4}$ in.
No. 296. Screwed for Iron ...	3/2	4/4	6/9	11/9
Screw Boss ...	3/6	4/10	7/6	12/9

"The Chelsea" Valve.

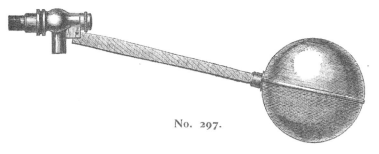

No. 297.

	$\frac{3}{8}$ in.	$\frac{1}{2}$ in.	$\frac{3}{4}$ in.	1 in.	$1\frac{1}{2}$ in.
No. 297. Screwed for Iron	2/5	2/11	4/1	6/6	11/11
Screw Boss ...	2/8	3/3	4/7	7/3	13/-

Equilibrium Valves

(For very High Pressure).

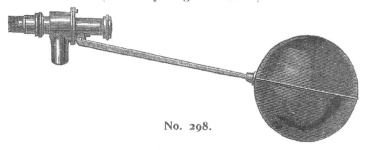

No. 298.

	$\frac{3}{8}$ in.	$\frac{1}{2}$ in.	$\frac{3}{4}$ in.	1 in.	$1\frac{1}{4}$ in.
No. 298. Screwed for Iron	3/4	3/10	5/-	8/3	13/8
Screw Boss	3/7	4/2	5/6	9/-	14/9

T Crapper & Co Traded from the following Chelsea Addresses

Address	Year	Status
42 Robert Street	1861-1866	Home and business.
52 Marlborough Road	1866-1930s	Brass foundry at rear.
50 & 52 Marlborough Road	1872-1885	Business and showrooms.
50, 52 & 54 Marlborough Road	1885-1930s	Business and showrooms.
120 King's Road	1907-1966 1963-1966	Business and showrooms, including penthouse flat used by Robert Gillingham Wharam. T Crapper & Co Ltd owned by John Bolding & Sons Ltd.

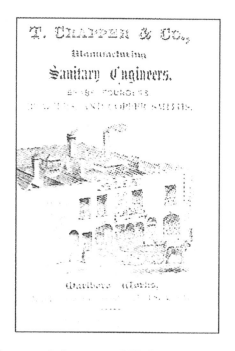

Lithograph image of T Crapper & Co.,
Marlboro' Works, Chelsea 1872

Thomas Crapper & Co., Marlboro' Works, Chelsea 1892

Photograph courtesy of Thomas Crapper & Company Ltd.

CHAPTER

MARLBORO' WORKS

Thomas Crapper's landmark premises were known as the Marlboro' Works, Chelsea, London. It is believed Thomas Crapper created the world's first bathroom showroom featuring large plate glass windows in which he was the first to prominently display sanitary ware, in particular many water closets, for which he became famous.

Crapper's first business premises in 1861 were at 42 Robert Street, Chelsea where he established himself as a plumber and brass finisher. He soon outgrew Robert Street and in 1866 acquired part of an existing brass foundry situated at the rear of 52 Marlborough Road, Chelsea. Thomas Crapper, plumber, became T Crapper & Co in 1866.

Surprisingly, Thomas Crapper was not the first plumber to occupy part of the Marlboro' Works. Simpson's Trade Directory of Chelsea 1863 confirms W H Chapman & Co, plumber occupied 54 Marlborough Road, however, his stay appears to have been short-lived.

The original Marlborough Works brass foundry was established in 1848 by John Martin, who in 1841 had a brass foundry in Sydney Terrace, Chelsea. Access to the foundry was gained through a wide archway between 52 and 54 Marlborough Road. The 1861 census for Chelsea enumerates Martin's foundry contained within a row of properties known as Buckingham Place. Martin operated the foundry until c1863 then it passed to Henry Wooldridge. From 1866 Wooldridge concentrated on brass finishing, retaining the workshop and the street frontage of 52 Marlborough Road.

Thomas Crapper took over part of the brass foundry in 1866, by 1871 he was employing 24 hands. From 1872 he operated the whole of the brass foundry including the brass finishing workshop together with the street frontage of 52 Marlborough Road. At, or about, the same time Henry Pike, a furniture dealer, vacated 50 Marlborough Road conveniently allowing Thomas Crapper to expand into 50 and 52 creating substantial sanitary ware showrooms

THOMAS CRAPPER & Co.,

Manufacturing Sanitary Engineers.

INTERIOR OF FINISHING SHOP.

MARLBOROUGH WORKS,

CHELSEA, LONDON, S.W.

Finishing Shop Marlboro' Works, Chelsea
Image from Thomas Crapper & Co Catalogue c1892-3

Courtesy of Thomas Crapper & Company Ltd.

complete with large windows facing onto the street. Thomas, with a hint of mischief, took pleasure in filling his window displays with a variety of water closets, an inspired marketing tactic which certainly got Crappers noticed. A lithograph of 1872 shows Crapper's Marlboro' Works incorporating 50 and 52 together with number 54 on the left hand side remaining as a typical house.

Thomas had to wait until c1885 for William Steel, a house painter and Alfred Wood, a carpenter, to vacate number 54 allowing him to remodel 50, 52 and 54 Marlborough Road as it appears in a photograph taken in 1892.

In addition to the foundry and sanitary ware showroom Thomas Crapper employed plumbers to fix items purchased, coppersmiths and braziers manufacturing all manner of plumbing items e.g. copper syphons for closet cisterns and bellows regulators for valve closets. Thomas was a leading merchant supplying sheet lead and lead pipe to plumbers in the surrounding district. Thomas had a lead shop where he manufactured lead syphons for flushing cisterns, lead valve boxes to house the flap valve of valve closets, soil pipes, ventilator cowls, traps and pumps, he also cast small ornamental lead items for rainwater hopper heads etc. It is estimated Thomas Crapper & Co employed around 100 workers during the 1890s.

It is evident from the 1892 photograph Thomas stocked a substantial range of sanitary ware including a variety of Bramah-style mechanical valve closets, plain and decorated wash down and wash out closets in pedestal form, basic pan and trap closets, a full range of cisterns including cast iron, ceramic and wood. He stocked wash basins and stands, plug basins, sinks, taps, waste fittings, urinals and baths. He also stocked copper, steel and cast iron pipe and fittings, salt glazed ware pipe and fittings including interceptor traps, grease traps, cast iron manhole covers, boilers, tanks, radiators, speaking tubes and spare parts. Thomas Crapper became not only the leading supplier of plumbing materials in London but also a leading distributor of specialist plumbing equipment throughout most of the country.

Trade directories are a valuable resource, they provide evidence of the diverse range of businesses in Marlborough Road as well as documenting Thomas Crapper & Co's rapid expansion.

Simpson's 1863 Trade Directory for Marlborough Road, Chelsea included:

➢ 50 Marlborough Road : Henry Pike, Broker
➢ 52 Marlborough Road : John Martin, Brass Founder

THOMAS CRAPPER & Co.

Marlboro' Works, Chelsea, S.W.

Birds Eye View of Marlboro' Works, Chelsea
Image from Thomas Crapper & Co Catalogue c1892-3

Courtesy of Thomas Crapper & Company Ltd.

> 52 Marlborough Road : A Fletcher, Carpenter
> 54 Marlborough Road : W H Chapman & Co, Plumber

There were 68 tradespeople listed residing at business premises in Marlborough Road in 1863. The various occupations included: E Humpherson; carpenter at 45, J Jones; watchmaker at 46, J W Snowden; furnishing undertaker at 48, C Wray; cow keeper at 51, G Turner; boot maker at 88, M Joseph; clothier at 94, Hezekiah Hobrough; grocer at 116, J Reed; eel pie house at 149 and Alexander Miles; silversmith and pawnbroker at 159 Marlborough Road.

> The Chelsea Trade Directory for 1878 describes T Crapper & Co, 50 & 52 Marlborough Road as lead merchants.
> The Post Office London Directory 1882 lists: Thomas Crapper & Co, 50 & 52 Marlborough Road, Chelsea; brass founder and lead merchant.
> The Business Directory of London 1884 mentions Thomas Crapper & Co, 50 & 52 Marlborough Road, Chelsea, brass founders and lead merchants.
> The Commercial Directory of 1895 states Thomas Crapper & Co, sanitary engineers etc., 50, 52 & 54 Marlborough Road, Chelsea, London.

Between 1866 and 1892 Thomas Crapper redeveloped his Marlboro' Works, he renovated the old buildings and extended his manufacturing capacity to the full extent of the site. His business was booming.

T Crapper & Co utilised coal fired boilers to generate steam power to operate the foundry's machinery, several chimneys protruding through the roofs, shown on the bird's-eye view of the works, indicate coal fires provided heating to the offices and showrooms. Gas lighting illuminated the offices and showrooms, whilst natural light from the large lantern windows was deemed sufficient elsewhere. Electricity was installed to the works during the 1890s and a telephone system was installed prior to 1900.

Depicted on each view of the Marlboro' Works 1872, 1892-3 and King's Road in 1908 is a horse and dray either entering or leaving the works. Edith Crapper, a great niece of Thomas Crapper, when visited by Wallace Reyburn seeking information for his book *Flushed with Pride, The Story of Thomas Crapper* (1969) remembered *'one of Crapper's carmen, as they used to call the men who made the deliveries on horse-drawn open drays, was a man named Lush ... with his dray laden with lavatory pans, cisterns and flush pipes.'*

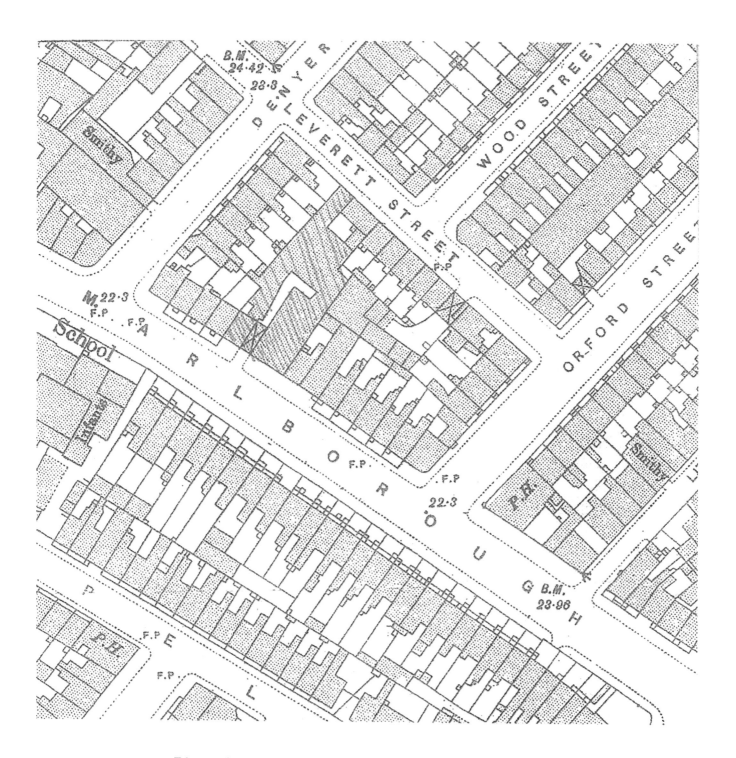

Plan of T Crapper & Co Marlboro Works 1894-96

Courtesy of Thomas Crapper & Company Ltd.

Crapper & Co probably owned 3 or 4 horses and drays, utilised for deliveries around London and its suburbs. Horses were a valuable asset to the business and were likely to be stabled in the rear yard of the works or perhaps nearby. Throughout Thomas Crapper's career horse drawn transportation of goods remained the most important means of delivery from manufacturer to customer. Thousands of working horses remained on the streets of London until the arrival of the commercial motor vehicle, prior to the outbreak of the First World War. For deliveries further afield Crappers either delivered their goods to the nearest railway goods yard or canal wharf e.g. Paddington, for delivery by barge via the national canal network. Goods ordered for export went direct to London Docks where they would be loaded aboard ships for destinations in Europe, America, Canada, Australia etc.

Plumbers employed by Thomas Crapper & Co normally walked to the job when working locally, the apprentice carrying the heavy bag of tools. Should an order for lead or cast iron pipework etc be required they would use a hand barrow or cart and think nothing of walking four or five miles to the job. The Author still has a vivid, painful memory from personal experience struggling to push uphill a large hand cart laden down with cast iron pipes and guttering whilst an apprentice plumber in the mid 1960s.

Thomas Crapper's historic Marlborough Works, Draycott Avenue, formerly Marlborough Road, Chelsea was sold for redevelopment c1936. Thomas Crapper & Co Ltd shrewdly retained the well-known Marlboro' Works name for their prestigious King's Road showrooms, offices and works premises. (see Chapter 11)

In 1948 Thomas Crapper & Co Ltd held a celebration centenary dinner in recognition of their famous Marlboro' Works, Chelsea, established by John Martin in 1848. No record of Crapper's actual centenary celebration of 1961 has been discovered.

THOMAS CRAPPER & Co.,

Patentees and
Manufacturers of Sanitary Appliances,

MARLBORO' WORKS, CHELSEA, S.W.

AIR-TIGHT MANHOLE COVERS.
AIR INLET VALVES.
Automatic Flushing Tanks. .
CAST IRON AND PORCELAIN BATHS.
Cast Iron Drain Pipes and Junctions.
ELASTIC VALVE CLOSETS.
ENAMELLED FIRE CLAY SINKS.
GREASE TRAPS.   GULLEY TRAPS.
Galvanized Iron Tanks and Cisterns.

The Improved Kenon Disconnecting Trap,

SUGGESTED BY

PROF. CORFIELD, M.A., M.D. (Oxon), F.R.C.P., Hon. A.R.I.B.A.

Improved Bath and Lavatory Fittings.
WASH-DOWN CLOSETS.
Registered White Glazed Blocks for Inspection Chambers.
SYPHON WATER WASTE PREVENTERS.

Catalogue and Price List,

July, 1895.

(SUBJECT TO ALTERATION WITHOUT NOTICE.)

ALL PREVIOUS ISSUES CANCELLED.

5

CHAPTER

CRAPPER CATALOGUES

Thomas Crapper was a brilliant exponent at advertising his products and services, in particular when promoting his brand name Thomas Crapper & Co which he registered as his trade mark in 1881. Crapper, of course, was not alone in promoting his business, his main competitors e.g. John Bolding, George Jennings and Dent & Hellyer were all adept in the art and practice of badging their own water closets, cisterns, basins, taps etc. The simple reason Thomas Crapper became so successful was a combination of his attention to detail, quality and service, fuelled by a booming economy and the relentless promotion of his personal status of holding a Royal Warrant of Appointment.

Thomas Crapper utilised trade directories, journals, architects' compendiums, news sheets etc to promote his company and products. Crapper quickly recognised the commercial benefit of producing his own trade catalogue which he issued to his building and plumbing customers, other potential clientele, particularly architects and engineers as well as government departments responsible for specifications and purchasing budgets.

Trade catalogues first appeared in the 18th century with Thomas Chippendale's *The Gentlemen and Cabinet Makers Directory* of 1754. Early sanitary ware advertisements took the form of single sheet posters or announcements in newspaper and periodical publications. Enoch Wood & Sons, potters of Burslem, Staffordshire published, in 1823, an illustrated advertisement for their extensive range of water closet pans including price list. From the 1840s sanitary ware manufacturers e.g. Doulton & Watts, James Stiff and Stephen Green all advertised in trade journals i.e. The Builder, The Plumber & Decorator and Gas & Sanitary Engineering. Publishers of educational books on plumbing and sanitary engineering recognised the potential revenue source and incorporated discretely placed trade advertisements in the rear pages of technical books.

Advertisement for The Plumber and Decorator Journal c1880

Typical Plumber's business card 1868

The Author's Gt. Gt. Grandfather was apprenticed, in 1850,
to his uncle Thomas Stevenson, Plumber, Glazier & Painter est. 1827

The Great Exhibition of 1851 was a major showcase for manufacturers leading to many firms discovering the value of producing their first trade catalogue. Elaborately illustrated business cards became the norm, these were handed out by large manufacturing concerns like Bramah & Co. Even the local plumber benefitted from having his own business card.

Industrialisation together with a rapid increase in the population during the Victorian period resulted in concentrated urbanisation. These events created tremendous opportunities for the likes of Thomas Crapper and other eager entrepreneurs, many of whom seized and benefitted from the challenge that lay before them.

Surviving trade catalogues, particularly pre-1900 are very rare, most having been lost or subject to the ravages of time. These trade catalogues were, after all, issued as day to day working reference books, subject not only to wear and tear but disposed of when outdated or superseded due to continuous advances in manufacturing, new patent inventions and the effects of legislation, particularly between 1870-1900.

Competition amongst manufacturers and merchants intensified culminating into who could produce the most elaborate trade catalogue. In the Author's opinion, two of the finest trade catalogues ever produced were by T W Twyford and J Dimmock & Co, both sanitary ware potters of Hanley, Staffordshire. Simon Kirby of Thomas Crapper & Co Ltd has a wonderful collection of historic trade catalogues including fine examples of Messrs Twyford & Dimmock.

Thomas Crapper was at the forefront in recognising the benefit not only of producing an excellent catalogue but also encouraging the passing public to view his bathroom showroom via his radical introduction of large plate glass windows at his Marlboro' Works. Crapper displayed the latest designs in plain and decorated sanitary ware for all to view, envy and aspire to own. Crapper's main competitor John Bolding & Sons opened their bathroom showroom at Davies Street, Mayfair, London in 1891.

Examples of Surviving Catalogues by Thomas Crapper & Co

1. T CRAPPER & CO., Manufacturing Sanitary Engineers, Brass Founders, Braziers and Coppersmiths, Marlboro' Works. A basic paperback catalogue 1872 for plumbers and brass founders. This catalogue includes utilitarian items e.g. polished mahogany toilet paper boxes.

Pedestal Wash=down Closets.

"The Improved Marlboro'."

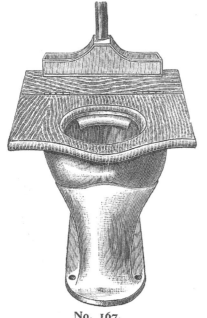

No. 167.

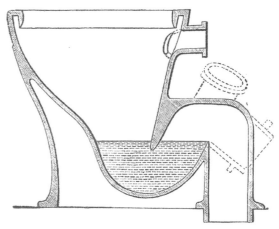

Section.

No. 167.	White Ware Basin and Trap, in one piece	...	...	£1 14 6
	Ditto, with Raised Ornamentation ...	...	Extra	5/-
	,, ,, ,, Tinted	...	,,	10/6
	Paper Box	...	...	,, 7/9
	Mahogany or Walnut Seat ...	...	,,	17/9

"The Deluge."

No. 168.

White or Ivory Basin, as shewn, with Raised
Ornamentation and Slop Top in one piece, Paper
Box and Mahogany or Walnut Seat with Flap
and Brackets

£5 8 6

No. 168.

60

2. THOMAS CRAPPER & CO., Patentees and Manufacturers of Sanitary Appliances, MARLBORO' WORKS, MARLBOROUGH ROAD, CHELSEA 1883. This catalogue was in essence Messrs Twyford's catalogue including T W Twyford's Patent CROWN and NATIONAL water closets. Twyford's established c1853 also manufactured ceramic items for the brewery trade e.g. wine and sherry spirit dispensers, beer pump handles etc. These items featured prominently in Crapper's 1883 catalogue and Twyford's 1889 catalogue.

3. THOMAS CRAPPER & CO., Patentees and Manufacturers of Sanitary Appliances, MARLBORO' WORKS, CHELSEA, 1886. This catalogue was Crapper's first known catalogue to feature his own named products e.g. Marlboro' Wash Out Closet and Chelsea Flush Down Closet. Crapper's own various elastic valve closets were illustrated noting that his No.1 was as supplied to Sandringham House, H.R.H. THE PRINCE OF WALES and Eastwell House, H.R.H. THE DUKE OF EDINBURGH.

4. THOMAS CRAPPER & CO. (presumed c1892-93) This undated catalogue most importantly includes illustrations of Crapper's Marlboro' Works external bird's eye view plus the only known internal illustration of Crapper's brass polishing workshop. Contents are similar to the 1886 catalogue.

5. THOMAS CRAPPER & CO., CHELSEA, LONDON. This catalogue and price list July 1895 is one of Thomas Crapper's finest catalogues with crimson cover and gold embossed feathers emblem of H.R.H. THE PRINCE OF WALES. The contents are the most comprehensive to date, the illustrations are sepia coloured.

6. THOMAS CRAPPER & CO., SANITARY ENGINEERS, BY APPOINTMENT TO HIS ROYAL HIGHNESS THE PRINCE OF WALES, CHELSEA, LONDON, SW 1897. This paperback catalogue features the Royal Coat of Arms on a light blue cover. Contents similar to the 1895 catalogue but not as impressive.

7. THOMAS CRAPPER & CO., SANITARY ENGINEERS, CHELSEA, LONDON, SW3. This catalogue and price list January 1899 with crimson cover and gold embossed Royal Coat of Arms. Comprehensive contents including Crapper's Valveless Water Waste Preventer Patent 4990, originally granted to Albert Giblin 9th April 1898.

8. THOMAS CRAPPER & CO., LONDON. This catalogue and price list 1902, with crimson cover and gold embossed Royal Coat of Arms. Inside

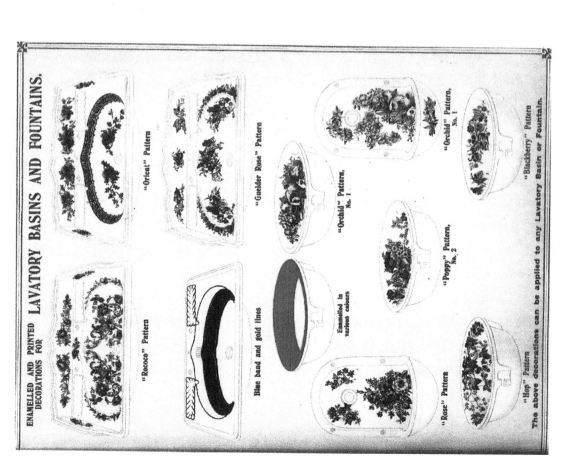

Examples of decorated sanitary ware supplied to Thomas Crapper & Co by Geo Howson & Sons, Hanley.

Images from Howson's 1900 Sanitary Appliances catalogue.

BY APPOINTMENT TO H.R.H. KING EDWARD VII AND H.R.H. THE PRINCE OF WALES. Comprehensive contents and illustrations.

9. THOMAS CRAPPER & CO LTD., CHELSEA, LONDON, SW3. BY APPOINTMENT TO HIS MAJESTY THE KING. Catalogue and price list No.28 February 1935. Crimson cover and gold embossed Royal Coat of Arms. Includes coloured bathroom suites and photographic images.

10. THOMAS CRAPPER & CO. LTD. Manufacturing Sanitary Engineers, 120 KING'S ROAD, CHELSEA, LONDON, SW3. BY APPOINTMENT TO THEIR LATE MAJESTIES KING EDWARD VII AND KING GEORGE V, hardback catalogue 1954, crimson cover and gold embossed Royal Coat of Arms. Comprehensive catalogue including coloured photographic images.

11. THOMAS CRAPPER & CO. LTD. This 1964 catalogue is very similar to the 1954 edition.

In 1891 Robert Marr Wharam produced a 31-page hardback book entitled *Hints on Sanitary Fittings and their Application.* Wharam's book is full of Thomas Crapper's designs and patents with illustrations from Crapper's catalogues. Almost certainly Wharam collaborated with Thomas Crapper to develop his book, the printer of Wharam's book also produced Crapper's catalogues. Robert M Wharam may have produced his book in support of his application to become a Member of the Royal Sanitary Institute. It is noticeable that no mention of Thomas Crapper appears in Wharam's book.

Many companies imitated Crapper's catalogues including John Bolding & Sons Ltd, their 1930 sanitary ware booklet, black matt cover with silver print featuring The Prince of Wales feathers emblem and By Appointment, is a cheap imitation of Thomas Crapper's 1895 catalogue.

Thomas Crapper & Co Ltd, Sanitary Appliances catalogue No.28, published in 1935, includes many products available in 1895. In 1935 Crapper's showrooms, offices, warehouses and works were situated at 120 King's Road and 47a King's Road, Chelsea. Coloured bathrooms were available in three shades of green, five blue, two pink, two primrose, black, amber, old ivory, lilac and putty. Coloured sanitary ware was developed and first introduced in 1925 by Armitage superseding elaborate, ornate, embossed and printed Victorian designs, which enjoyed a resurgence in the 1980s.

Automatic Flush Tanks,

For keeping Drains constantly and effectually flushed.

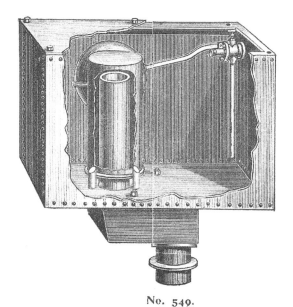

No. 549.

No. 549.

Galvanized Iron Flush Tank, with Trapping Box, Strong Zinc Cylinder, and Regulating Supply Valve, complete

			£	s.	d.
10 Gall.	...	...	£2	7	6
15 ,,	...	...	2	11	6
20 ,,	...	...	2	15	6
25 ,,	...	...	2	19	0
30 ,,	...	...	3	4	6
40 ,,	...	...	3	12	6
50 ,,	...	...	4	2	6
60 ,,	...	...	4	12	6
75 ,,	...	...	5	6	6
100 ,,	...	...	5	19	6
150 ,,	...	...	7	9	6

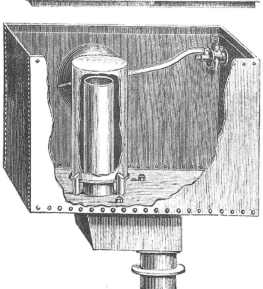

No. 550.

No. 550.

Galvanized Iron Flush Tank, with Trapping Box, Strong Zinc Cylinder, and Regulating Supply Valve, complete

			£	s.	d.
10 Gall.	...	...	£2	7	6
15 ,,	...	...	2	11	6
20 ,,	•••	•••	2	15	6
25 ,,	•••	...	2	19	0
30 ,,	...	...	3	4	6
40 ,,	...	...	3	12	6
50 ,,	...	...	4	2	6
60 ,,	...	...	4	12	6
75 ,,	...	...	5	6	6
100 ,,	...	...	5	19	6
150 ,,	...	...	7	9	6

If with Copper Cylinder   *Extra* 7/6

If without Regulating Supply Valve   *less* 8/6

The final listing in the 1935 catalogue emphasised that Thomas Crapper & Company Ltd *'having their Works at the rear of their Warehouses and Showrooms, and a Staff of skilled workmen, possess exceptional facilities for executing with speed and efficiency alterations and repairs connected with Valve Closets, Water Waste Preventers, Automatic Flushing Tanks, Baths and Lavatories ...'* The company's foundry produced castings in gun metal, brass, copper and lead.

Sandringham House, Norfolk

Photograph courtesy of www.TourNorfolk.co.uk

Parish Church of St Mary Magdalene, Sandringham

Photograph courtesy of © Philip Halling

6

CHAPTER

SANDRINGHAM CONNECTION

The Thomas Crapper connection with Sandringham is usually quoted as 1886. The Author's research has uncovered Crapper family connections from 1841 and possibly back to 1827.

Queen Victoria purchased the Sandringham Hall Estate, comprising 7,000 acres in 1862 for £220,000, as a gift for her son and heir Albert Edward, Prince of Wales. The hall, a Georgian mansion c1750, was owned by The Hon. Charles Spencer Cowper, stepson of Viscount Palmerston the Prime Minister. Cowper only occasionally visited the hall preferring to live in Paris. Some improvements were made to the estate during the 1850s, however the hall was allowed to slowly deteriorate.

The Prince of Wales made the hall suitably habitable and in 1863 took up residence with his wife of three weeks, Princess Alexandra of Denmark. Unhappy with the hall's proportions The Prince had it demolished in 1867 and a new Sandringham House, built by Goggs of Swaffham, was completed by 1870.

During my research I came upon a publication *'From Country House to Royal Retreat'* produced for an Exhibition on Royal Norfolk 2012. Contained within a paragraph referring to the rebuilding of Sandringham House taking nearly 3 years, completed in 1870, states *'The new house had all the latest inventions including the installation of modern cisterns under the supervision of Thomas Crapper.'* Being surprised to discover Crapper's involvement at such an early date (1867-70), further enquiries found no evidence to support this statement, concluding Thomas Crapper's name was included because he was so well known.

The Prince of Wales succumbed to typhoid fever in 1871 and The Queen, anxious for his life, visited him at Sandringham. Victoria lost her beloved husband Prince Albert ten years earlier to the same, often fatal disease. The Prince of Wales eventually made a full recovery, however a very public inquest

Elastic Valve Closets.

The Overflow is carried into Conductor at back of Discharge Valve by which means the Overflow is kept clear. A small pipe is fitted from Supply Arm to Trap of Overflow by which the latter is always kept supplied with water, thus preventing syphonage.

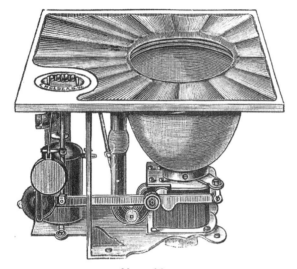

No. 188.

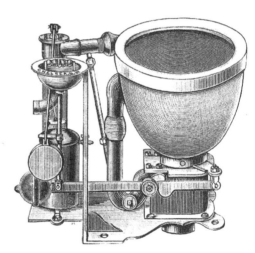

No. 189.

No. 188. Valve Closet, with White Ware Flushing-rim Basin, Large Slop Top and Dish in one piece, 1 in. Supply Valve, Copper Air Regulator, complete as shewn £4 14 6

„ 189. Ditto   ditto   ditto   without Slop Top 3 9 6

If with 1¼ in. Valve .... Extra	3/6
„ Ornamental Basin .... „	3/6
„ White and Gold Basin .... „	8/9
„ Box Enamelled inside .... „	4/9
„ Box fitted with Brass Top .... „	6/3
„ Box fitted with Union to connect Ventilating Pipe .... „	3/9
„ 4-in. Outlet .... „	7/9

ensued, with the opinions of several esteemed sanitary engineers pontificating as to the cause of The Prince's near fatal illness.

S Stevens Hellyer in a magazine article suggested The Prince of Wales' illness was *'perhaps from bad plumbing.'* This, as they say, put the cat amongst the pigeons! William Paton Buchan in his book *Plumbing* (1883) recalls *'about the year 1871 great excitement was caused in the sanitary world by the severe attack of typhoid fever which His Royal Highness The Prince of Wales then had. Londesborough House* (the home of the Countess of Londesborough where, it was thought, The Prince caught typhoid) *was turned inside out so far as its drains and sanitary fittings were concerned ... I was astonished to read the following statement in connection with the plumbers' work at Sandringham House in Cassell's Technical Educator ... the space between the bottom of the valve and water in an ordinary valve closet has a small air pipe running into the ventilating soil pipe. We may mention Sandringham House as an instance of the adoption of this precaution, for non practical readers the installation so described would allow sewage gas from the soil pipe to come back through the air pipe and poison the air where the closet is situated.'* Buchan reported *'upon the authority of the firm who executed the work, the above description was a mistake and The Technical Educator have taken means to rectify the error.'*

Buchan wrote to His Royal Highness The Prince of Wales regarding his illness and received a diplomatic reply stating *'The Prince was quite unaware in what manner he had caught his typhoid fever.'* From Cassell's description we can be sure Bramah-style valve closets were installed at Sandringham in 1870, however as a result of substantial publicity a question mark hung over the plumbing installations at Sandringham House for some time.

The Prince added a ballroom to Sandringham House in 1881, however by 1885 it became obvious that Sandringham's general sanitary provision had become outdated and the drains rapidly deteriorating. The Prince of Wales requested Mr Rogers Field, the senior partner in a London firm of civil engineers, who specialised in hydraulics and drainage, to urgently draw up a scheme of improvements at Sandringham House.

An article appeared in the Bristol Mercury on the 1st February 1886 entitled *SORE THROAT AT SANDRINGHAM ... 'We cannot suppose that in the house at Sandringham insanitary conditions have been permitted to prevail, for it was here the Prince was so seriously ill from typhoid, by which attention should*

Independent Spray and Shower Bath.

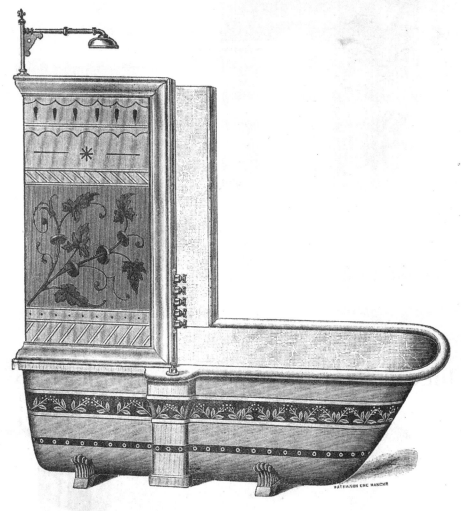

No 2

No. 2. 6ft. Enamelled Cast Iron Bath, with Zinc Spray Enclosure, Enamel-painted inside and the Whole Decorated outside, Nickel-plated Hot and Cold Valves, including Controlling Valves for Spray and Shower, and with Lift-up Waste, complete as shewn **£27 17 6**

  If with Copper Spray Enclosure ... Extra £4 15 0

have been forcibly directed to the sanitation of Royal residences. The Princess of Wales has been subjected to risks and many distinguished visitors who have been at Sandringham since Christmas ... Any gentleman would regard it as a paramount duty to his family to select his home in a district where sanitary statistics were satisfactory, or he would set about at once to remedy the defects as far as practicable. ... The Prince of Wales can scarcely, with his many duties, be expected either to understand or attend to such matters, but his advisors should.'

On the 18th February 1886 a 42-page bound report on Sandringham drains and sanitation was delivered to the Prince, who immediately gave instructions to proceed with all haste.

Mr Rogers Field designed and supervised the construction of water supplies, drainage, sewage disposal arrangements to hospitals, asylums, schools and private residences throughout the country. He was the vice president of the Sanitary Institute, with a number of patented inventions to his name including a valve closet, an intercepting trap and an automatic flushing cistern. Field was fastidious in his work, he was a life-long bachelor who lived the whole of his life in the house where he was born, together with his four spinster sisters. Field lived in style at Squire's Mount, Hampstead, employing several servants and providing a lodge house for his gardener. At his death in 1900 he left an estate of £65,500 (c£6.5m in 2014).

Rogers Field selected and instructed tradespeople and suppliers he had worked with before, who he could trust and rely on especially for such a prestigious commission. Rogers Field instructed Thomas Crapper who was delighted to be of service to HRH The Prince of Wales. He supplied over thirty water closets, cisterns, basins, baths, urinals, sinks, disconnecting traps, flushing tanks and inspection chamber covers. Most of the products he supplied carried Thomas Crapper's name or logo T.C. & Co.

Thomas Crapper ensured his royal contract was completed by the end of 1886 and fulfilled to The Prince of Wales' entire satisfaction. Within months of Thomas completing the Sandringham commission he contracted a serious case of smallpox which nearly claimed his life. Some time later Thomas joined a London Masonic lodge, no doubt engaging in charitable fund raising whilst forming valuable contacts beneficial to his business activities. Several members of the Royal Family are known to have been involved with Freemasonry.

Combined Cabinet Lavatories.

No. 52.

No. 52. Pitch-pine Cabinet Stand, 2ft. 9in. by 2ft. with 1¼in. polished Sicilian Marble Top, ¾in. Skirting, 4in. high, 14in. inside diameter, Marbled Basin, Grating and Stand Pipe, with Pull lettered Waste, and Hot and Cold Supply Valves £8 2 6

,, 53. Mahogany ditto ditto ... 8 14 6

St. Ann's Marble Top, Extra £1 15 0

Oval Basin   ,,   1 6 6

If Grating, Waste Pull and Valves plated   Extra 9/6

No. 54. Mahogany Cabinet Stand, 4ft. by 2ft., with Tiled Back, 1-in. Sicilian Marble Slab. 14in. White Ware Tip-up Basin and Receiver, and two Cam-action Valves, lettered Hot and Cold £10 17 6

,, 55. Ditto, ditto, with 16in. Tip-up Basin and Receiver... 11 7 6

Marbled Basin 14in. extra over white 0 2 6

Ditto   16in.   ditto   0 3 6

No. 54.

So, which member of Thomas Crapper's family knew more about the Sandringham Estate than he did? Well – it was none other than his wife Maria Crapper nee Green. Maria was born at Hemsby in Norfolk and by the age of 12 she was, like many young girls, put into service. Maria was employed as a house servant by the Reverend George Browne Moxon, rector of Sandringham, Norfolk. Maria, age 15, is enumerated in the 1851 census for Sandringham living at The Rectory, the home of George Browne Moxon, rector age 56, Bertha his wife 37 and house servant Abigail Green 43.

Abigail Green was the mother of Matilda Crapper, wife of Thomas's brother George. Abigail appears in the 1841 census at Sandringham and was probably in service there from 1827 when the Reverend George Browne Moxon became rector of Sandringham. Matilda was illegitimate and raised by her grandparents at Scratby, Norfolk. In 1851 Matilda is living with her aunt Sarah Crapper and family in Thorne, Yorkshire. Robert Crapper age 10, brother of Thomas, is recorded in the 1841 census residing in the household of Sarah Crapper's brother Robert Green, farmer at Hemsby, Norfolk alongside Maria Green, age 5, future wife of Thomas Crapper.

The Rectory was a substantial residence positioned between Sandringham House and St Mary Magdalene Church. Abigail Green left The Rectory when she married William Chambers of Dersingham in 1854. Abigail died a few years later and was buried in the churchyard of St Mary Magdalene, Sandringham on 29th May 1859. Interred in the same churchyard are Prince Alexander John of Wales, who died in 1871, son of King Edward VII, Prince John who died in 1919, son of King George V and John Spencer who died in 1960 brother of Princess Diana. I wonder if Matilda, George and Thomas Crapper attended Abigail's Sandringham funeral, it is likely that Maria Green, who was close to her aunt, would have been present.

It is thought Maria remained in service at Sandringham until she married Thomas Crapper in 1860. It is probable Matilda Crapper had a hand in bringing Thomas and Maria together, a match that lasted until Maria's death in 1902.

The Reverend George Browne Moxon served as rector of Sandringham for 37 years, he died there in 1866. He witnessed the arrival of the new owner of Sandringham Hall in 1862 – HRH The Prince of Wales. It would be pleasing to think Thomas Crapper escorted Maria, a lady of substance, on a return visit to Sandringham where he could proudly show her his company's work.

Hot Water or Steam Radiators

For warming Public Buildings, Mansions, Churches, &c.

These Radiators being Ornamental do not require Cases or Coverings.

(Also made in other Designs.)

No. 290.

Single. (7 in. wide.)

	Height	2 ft.			2 ft. 3 in.			2 ft. 6 in.			2 ft. 9 in.			3 ft.			3 ft. 3 in.		
		£	s.	d.	£	s.	d.	£	s.	d.	£	s.	d.	£	s.	d.	£	s.	d.
Length	2 ft. 0 in.	3	15	0	3	18	0	4	1	0	4	4	0	4	7	0	4	10	6
,,	2 6	4	14	0	4	17	0	5	0	0	5	3	0	5	6	0	5	9	6
,,	3 0	5	6	0	5	9	0	5	15	0	5	18	0	6	1	0	6	5	0
,,	3 6	6	5	0	6	8	0	6	14	0	6	17	0	7	3	0	7	10	0
,,	4 0	7	6	0	7	9	0	7	15	0	8	2	0	8	8	0	8	15	0
,,	4 6	8	9	0	8	15	0	9	1	0	9	7	0	9	13	0	10	0	0
,,	5 0	9	7	6	9	13	6	10	0	0	10	10	0	11	0	0	11	8	0
,,	6 0	11	5	0	11	11	0	12	0	0	12	10	0	13	0	0	13	13	0

Double. (11 in. wide.)

	Height	2 ft.			2 ft. 3 in.			2 ft. 6 in.			2 ft. 9 in.			3 ft.			3 ft. 3 in.		
		£	s.	d.	£	s.	d.	£	s.	d.	£	s.	d.	£	s.	d.	£	s.	d.
Length	2 ft. 0 in.	7	16	0	8	2	6	8	8	0	8	14	0	9	1	0	9	7	6
,,	2 6	9	14	0	10	0	0	10	6	0	10	12	0	10	18	0	11	11	0
,,	3 0	10	19	0	11	8	0	11	18	0	12	4	0	12	10	0	12	16	0
,,	3 6	13	9	0	14	2	0	14	15	0	15	7	0	15	19	0	16	5	0
,,	4 0	15	13	0	16	6	0	16	18	0	17	16	0	18	15	0	19	8	0
,,	4 6	17	10	0	18	3	0	18	15	0	19	14	0	20	13	0	21	5	0
,,	5 0	19	8	0	20	0	0	20	13	0	21	11	0	22	10	0	23	9	0
,,	6 0	21	5	0	21	18	0	22	17	0	23	9	0	24	7	0	25	6	0

Thomas Crapper's involvement with Sandringham continued when, a few years later, he was further called upon to provide the sanitary installations to Park House (where Princess Diana was born in 1961). A disastrous fire occurred at Sandringham House in 1891, badly damaging the top floor, which had to be completely rebuilt, further guest accommodation was added in 1892. Thomas Crapper & Co Ltd continued to supply sanitary fittings to the Royal Family at Sandringham House long after the death of Thomas Crapper. A few original Thomas Crapper & Co sanitary ware fittings and inspection chamber covers remain in situ at Sandringham.

THOMAS CRAPPER & COMPANY,

SANITARY ENGINEERS,

BY APPOINTMENT TO H.R.H. THE PRINCE OF WALES.

SHOW ROOMS AND OFFICES:

50, 52, & 54, MARLBOROUGH ROAD,

CHELSEA, LONDON, S.W.

CHAPTER

BY APPOINTMENT

The proudest moment of Thomas Crapper's career was the day he became the holder of a Royal Warrant of Appointment, entitling him to display the Royal Coat of Arms on his business premises and stationery in recognition of his service, quality and excellence.

Thomas Crapper & Company's 1895 catalogue, produced with crimson covers, prominently displaying The Prince of Wales feathers emblem embossed in gold, leaves no doubt in customers' minds as to his acclaimed prestige, much to the envy of his competitors. The catalogue's inside cover boasts *'THOMAS CRAPPER & COMPANY, SANITARY ENGINEERS, BY APPOINTMENT TO H.R.H. THE PRINCE OF WALES.'*

Thomas Crapper received his first Royal Warrant c1886, granted for regularly supplying goods and services of the highest standard to the Royal Household of HRH The Prince of Wales. Thomas Crapper & Co supplied and installed sanitation over many years to Sandringham House, Buckingham Palace and Windsor Castle. Other prestigious state contracts included Westminster Abbey.

The earliest reward for royal service came in the form of a Royal Charter granted to the Weavers' Guild in 1155 by Henry II. Trade guilds were later known as Livery Companies. The Worshipful Company of Plumbers obtained its Grant of Ordinances in 1365.

Royal recognition of plumbers dates back to the reign of King Edward III who established the office of 'Plummer of the Kings Works' appointing Robert Horewode 'Sergeant of the Household'. The office of Sergeant Plumber to the King or Queen continued until 1782 when Burke's Act abolished the office. Henry VIII appointed tradesmen to 'Serve the Court' but it was the latter part of the 18th century when 'Appointed Tradesmen' started displaying the Royal Coat of Arms on their premises and stationery.

Stephen Hawkins' 'Pan Closet' 1821-1829
Thomas Minton made ceramic closet pans for Hawkins.

Josiah Wedgwood made this Bramah style 'Valve Closet' basin
for F G Underhay c1870.

In 1790 Joseph Bramah, water closet manufacturer, featured the Royal Coat of Arms when advertising in the Bristol Mercury. His main competitor, Robert Hardcastle, promoted the fact he was closet maker to their Royal Highnesses The Dukes of York and Clarence. Stephen Hawkins, in 1825, printed inside his highly decorated water closet pans *'By His Majesty's Royal Letter Patent Granted to S Hawkins for an Improvement upon Water Closets'* he also included the Royal Coat of Arms.

Mischievously Edward Johns' pottery at Armitage, Staffordshire, who did not hold a Royal Warrant, printed the Royal Coat of Arms emblem and *'Not by Appointment to the King'* on a Dolphin water closet pan. George Jennings advertised himself as *'The Greatest Name in Sanitation'* however it was Thomas Crapper who achieved his Royal Warrant of Appointment some 15 years before George Jennings & Company attained their first Royal Warrant in 1901.

When Edward VII ascended the throne in 1901, on the death of Queen Victoria, Thomas Crapper & Co were granted two further Royal Warrants: By Appointment to His Majesty the King and By Appointment to HRH The Prince of Wales (George V). A fourth Royal Warrant was granted to Thomas Crapper & Co Ltd in 1911, By Appointment to His Majesty The King - George V.

A further two Royal Warrants were granted to Thomas Crapper & Co Ltd by the Department of His Majesty's Privy Purse in 1938, By Appointment to the late King George V and later by the Department of Her Majesty's Privy Purse, By Appointment to the late King George V 1953. T Crapper & Co Ltd renewed their Royal Warrants every year until the end of 1963 when T Crapper & Co Ltd was sold to John Bolding & Sons Ltd as a result Thomas Crapper's Royal Warrants were assumed surrendered after 77 years of royal service. Royal Warrants are not transferrable and remain a valuable asset for any business fortunate to be granted 'By Appointment'.

The Royal Warrant Holders' Association was established in 1840, representing those holding Royal Warrants by Appointment. Currently there are approximately 800 Royal Warrant Holders. Tim Heald's book *A Peerage for Trade* is an interesting record of royal tradesmen spanning several hundred years.

There are, at the time of writing, 2,590 Peers of the Realm including 830 sitting in The House of Lords. Clear evidence that a tradesman's Royal Warrant by Appointment is less common than an unelected grace and favour

Spray Bath with Enclosure.

No. 1.

No. 1. 6ft. Cast Iron Bath, Enamelled inside, with Galvanized Supply and Waste Pipes, Spray Enclosure with Shower and Douche in Copper, Combined Hot and Cold Nickel-plated Supply Valves; and with Best-made Mahogany or Walnut Enclosure and Hand-painted Tiled-back Skirting, complete as shewn **£69 12 6**

lordship, often granted by politicians to each other as a reward for speaking up or perhaps keeping quiet!

In 1925 Robert M Wharam of Thomas Crapper, sanitary engineers, had the great honour of becoming the 31st president of the Royal Warrant Holders Association. His name is inscribed in gold lettering on The Honours Board of distinguished past presidents which hangs in The Council Chamber at No.1 Buckingham Place, London.

Crapper's Patent Seat=action Flush=down W.C.

Adapted for

Hotels

and

Public Buildings,

being

Automatic

in Action.

Commended by

the Trade Press

as being

" A very excellent

arrangement,"

" Simple and

effective."

No. 164.

No. 164.   Registered Flush-down W.C. Basin with Raised Ornamentation, 3-gallon Syphon Water Waste Preventer, and Polished Mahogany or Walnut Seat with Patent Automatic arrangement £6 17 6

If Raised Ornamentation on Basin is Decorated   Extra 10/6

CHAPTER

PATENTS AND DESIGNS

Thomas Crapper commenced his plumbing career one year after the Great Exhibition of 1851 held at Crystal Palace in London. The exhibition was a catalyst for new inventions, notably sanitation and, in particular, the flushing water closet. The opportunity to present to the world the latest innovations, designs and advances in manufacturing to hundreds of thousands of visitors and potential customers was irresistible to the likes of George Jennings, Guest & Chrimes, F. G Underhay, Stephen Green, John Ridgway and Josiah Wedgwood, to name just a few.

Joseph Bramah, a carpenter, and Yorkshireman like Thomas Crapper, manufactured the world's first commercially successful water closet, a mechanical valve closet, for which he was granted a patent in 1778. Bramah sold around 10,000 of his closets by 1800, even at the privileged sum of 7 guineas each plus installation demand was high. From 1780 Bramah incorporated Wedgwood's Queen's Ware pans, made to his own design, into his water closets which were regarded as the best for first class houses. Bramah's main competition came from Robert Hardcastle's less complicated mechanical pan closet, which was cheaper and more robust. The pan closet and the Bramah remained popular for about 100 years, with the pan closet being the choice of many plumbers who portrayed it on their business cards and invoices.

The first common water closets were made from sheet lead knocked up by plumbers into lead hopper pans with 'D' traps. Various pans for both valve and pan closets were made in lead complete with a valve service box or container if a pan closet, plus a lead safe tray. The plumber fabricated timber lead-lined cisterns which supplied water for the house requirements. Early piped water supplies from private water companies were intermittent, available for just one or two hours only per day. The plumber installed lead pipework to supply the water closet plus the lead soil discharge pipe from the closet to a convenient drain, if one existed. More often than not it went to a cesspit in the

Development of the Water Closet
1778-1878

Joseph Bramah
'Valve Closet' 1778

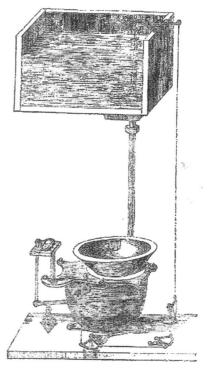

Robert Hardcastle
'Pan Closet' 1788

Thomas Smith
'The English Closet' 1844

George Jennings
'Monkey Closet' 1852

Stephen Green
'Improved Syphon Closet' 1852

John Roe
'Ped Wash Down Closet' 1878

rear garden or even directly into the cellar beneath the house, when it would be periodically manually emptied with buckets.

A plumber's work included the installation of wires, cranks and levers necessary to operate mechanical water closets, he also fitted the enclosing wooden surrounds. Thomas Crapper would have installed hundreds of Bramah and pan closets, particularly in affluent Chelsea and Kensington, it is likely he made a good living unblocking jammed valve closets, leaking joints, slack wires and damaged pans, not forgetting frozen and burst pipes.

The first all ceramic one-piece pan and trap, known as 'The English Closet', attributed to Thomas Smith was introduced in 1844. Stephen Green's all ceramic pedestal wash down water closet along with George Jennings wash out 'Monkey Closet' were introduced in 1852. John Roe produced a modern style pedestal wash down water closet including flushing rim in 1878.

The first known water waste preventer (flushing cistern) known as the 'Soil Pan Service Box', used for common water closets, was designed and registered to Stock & Sons of Birmingham in 1851. Messrs Stocks' advertisement in The Builder is detailed in its description, announcing *'The object of this contrivance is to facilitate the introduction of sanitary improvements into the dwellings of the working classes by the adoption of such an arrangement of cheap soil pan apparatus as will enable the constant supply system of water service to be made available for the removal of refuse by preventing the enormous waste of water in places of this description, to which water companies now so strongly and so properly object. By the use of this service box an ample supply of water is obtained, without the possibility of it being wasted by propping the valve, weighting the handle or by other means. It is so constructed that as long as the discharge valve remains open a further supply cannot enter the service box from the main.'* Stock & Sons' WWP was available 20 years prior to Parliamentary legislation.

The Industrial Revolution combined with the land grab of enclosure, the epitome of the ruling landed gentry, caused a great migration of labour from the country to urban living. Workers in the new manufactories were housed with their families in jerry-built, insanitary accommodation, hurriedly erected by speculative, unscrupulous landlords. Serious overcrowding resulted in major health issues forcing Parliament to introduce an Act in 1848 requiring a *'water closet, privy or ash pit to be provided for all new built and modernised houses.'*

Patent Disconnecting Trap.

(No. 10,332.)

Registered Design, No. 105,149.

Rd. Trade Mark No. 81,187.

"The Improved Kenon, Thomas Crapper & Co."

Advantages :—Provision at upper part of Trap for discharging into sewer any accumulation caused by accidental stoppage.

Easy access to passage for sweeping purposes, by means of a suitable brass cap with screw.

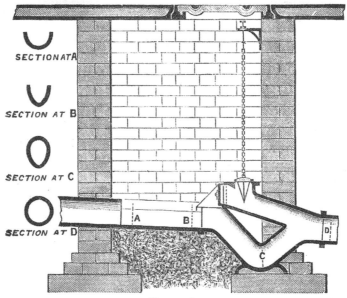

No. 506.

	4 in.	4 to 6 in.	6 in.	9 in.	12 in.
No. 506. **The Improved Kenon Trap,** No. 1 Pattern, of Glazed Stoneware with gun metal valve, galvanized iron chain, pull and bracket, and brass screw cap	29/6	31/-	32/6	46/6	89/6

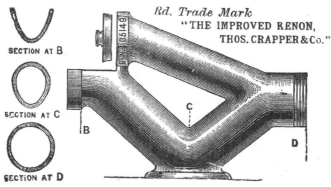

Rd. Trade Mark

"THE IMPROVED KENON, THOS. CRAPPER & Co."

No. 507.

	4 in.	4 to 6 in.	6 in.	9 in.
No. 507. **The Improved Kenon Trap,** No. 2 Pattern, of Glazed Stoneware	8/6	9/6	10/6	22/9

If Traps glazed white inside   Extra 4 in. 3/6 ; 4 to 6 in. 3/9 ; 6 in. 4/- ; 9 in. 7/6

,,   ,,   white inside and out ,,   7/- ;   7/6 ;   8/- ;   15/-

The raising of standards exposed a deficient national sewage disposal system, exacerbated by the massive surge in demand for water closets.

Dr John Snow identified the link between cholera and drinking water contaminated with sewage. A serious cholera epidemic occurred in a densely overcrowded area of London in 1854. Thomas Crapper escaped the disease, however many thousands of London residents were carried off by cholera and typhoid.

Great engineers such as Edwin Chadwick, George Roe and Joseph Bazalgette designed and supervised the installation of drainage systems in London and elsewhere, gradually improving the nation's health and hygiene. The Metropolitan Water Act 1871 required *'every water closet hereafter fitted to have an efficient water waste preventing apparatus.'* Thomas Crapper, an industrious, enterprising plumber recognised and seized the opportunity, quickly expanding into manufacturing his own mechanical valve water closets together with numerous other items to satisfy the huge demand for plumbing equipment of every description. Thomas's practical background experience combined with his exceptional technical ability enabled him to invent useful sanitary equipment and improve existing patented products e.g. the water waste preventer, drainage and ventilation fittings.

Thomas Crapper's reputation as a sanitary engineer and inventor was praised when The British Medical Journal on the 20th December 1890 gave almost a complete page, No.1433, to a report and analyses of new inventions and improvements in sanitary appliances. The report was based on a British Medical Journal inspection of appliances made by Thomas Crapper & Co, Marlborough Road, Chelsea. The following are extracts from the report;

'The Improved Kenon Disconnecting Trap (with illustration) – The special feature of this trap is that the pipe holding the trapping water is made egg-shaped instead of circular. The flow of sewage through it is consequently accelerated, it is more self-cleansing. They are absolutely necessary to break the direct connection between private drains and public sewers, in Messrs Crapper's improved form the disadvantages are likely to be reduced to a minimum.

Improved Automatic Flush Tank (with illustration) – With an annular siphon for rapid discharge of water, including a reversed action ball-valve preventing dribbling and continuous action which are the failings of many flush tanks now in the market.

Registered White Glazed Blocks
For Inspection Chamber Floors.
(No. 245,045.)

These Blocks have been designed to meet the objections usually raised by Engineers and Surveyors to the formation of Cement Channelling, a smooth and impervious surface being secured.

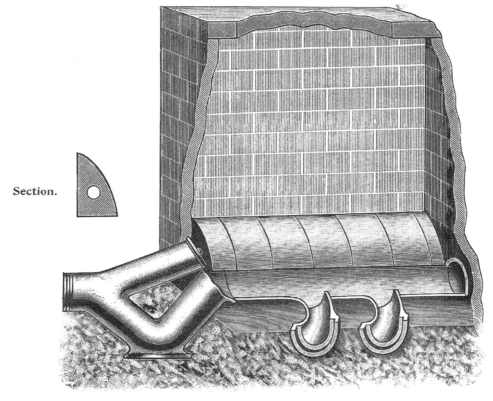

Section.

No. 528.

			3 in.	6 in.	9 in.
No. 528.	White Glazed Block ...	...	1/9	2/6	3/6

Cast and Wrought Step Irons
For Inspection Chambers.

No. 529.   No. 530.

		Plain.	Galvanized.			Plain.	Galvanized.
No. 529.	Cast Iron	1/-	1/8	No. 530.	Wrought Iron	1/6	2/3

Improved Grease Trap with Airtight Cover (with illustration) – *This gulley is connected with a discharge pipe from an automatic flush tank. The trap of the gulley is egg-shaped in section to secure its cleansing properly. This form of flushing grease gulley is a great improvement on the old forms of grease trap where the grease has to be removed by hand.*

Improved Siphon Action Water Waste Preventer – This is a small cistern to hold two or three gallons of water for flushing hopper or wash-out water closets. It is fitted with a siphon for rapid discharge of water on pulling the chain and can be fixed to the wall without brackets, the union supply pipe to the closet with the cistern is also out of sight. The cistern has therefore a more finished and ornamental appearance than the usual water waste preventer with its brackets for support and unsightly unions.

Improved Flush-down Water Closet – This is a pedestal short hopper closet of china with flushing rim. The trap beneath the basin is egg-shaped in section.'

Thomas Crapper's growing esteem as an inventor and designer is evident from the following testimonials:

The Parkes Museum of Hygiene, London, which opened in 1879, developed a permanent exhibition of sanitation. T Crapper & Co's Disconnecting Drain Trap was added to the exhibition, noting – *'specially suitable for deep manholes, the cross section of this trap is self cleaning.'*

House Drainage for Architects & Building Inspectors by G A T Middleton 1892, states *'The most important thing with this disconnecting chamber is undoubtedly the intercepting trap itself, of this there are very many first rate forms in the market, prominent among these being .... Crappers – Kenon.'*

Many of Thomas Crapper's inventions and designs were commended in leading publications e.g. The Architect 1887 and 1890, The Builder 1887, 1890 and 1892, The Plumber & Decorator 1889, Decorators' Gazette, Plumber & Gasfitters' Review 1889, The Building News 1890.

By 1895 Crapper had dropped the discredited, insanitary, mechanical old pan closet from his catalogue which also highlighted a move away from the declining popularity of the wash out closet in favour of the rising demand for the superior wash down closet. Thomas retained a selection of eight valve closets for his clients who perceived added value in the long established Bramah-style water closets. Crapper constantly strived to improve the valve

Improvements in Flushing Cisterns

Inventor:	Applicant:	Publication info:	Priority date:
GIBLIN ALBERT	GIBLIN ALBERT	GB189804990 (A) 1898-04-09	1898-03-01

4990. Giblin, A. March 1. Water-waste preventers, siphon-discharge. Improvements are described whereby an effective discharge can be obtained when the cistern is in any state between about half-full and full. In the chamber b connected with the siphon a is a projecting plate $f^{<1>}$ which makes joint with the plate d during the upward motion of the latter, but leaves a free passage for the water when the plate has reached the top of its stroke. The top of the chamber b is dished, so that the water collecting on it may prevent the entry of air round the rod c.

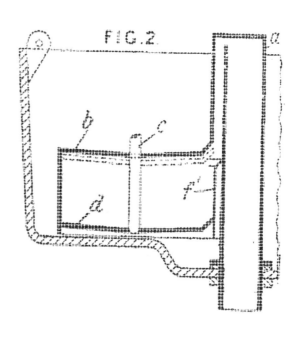

Albert Giblin

Photograph courtesy of Thomas Crapper & Company Ltd.

Giblin's original 1898 patent 4990 developed and improved by Thomas Crapper, featured in T Crapper's January 1899 catalogue closet e.g. he added a supply pipe between the flush pipe and the closet pan overflow trap, which if allowed to dry out lost its water seal resulting in foul air escaping from the valve box entering the room. Crapper named many of his water closets after Chelsea street names e.g. Marlboro, Ovington, Walton, Cadogan and The Chelsea.

Thomas Crapper offered his enlightened clientele a bespoke service, where he was happy to accommodate unusual requests e.g. his carpenter constructed a purpose-made blue upholstered closet chair with arms for Lillie Langtry, believed to be at the request of HRH The Prince of Wales.

Crapper's Patent 4990 WWP

Mr Albert Giblin was originally granted patent 4990 on 9th April 1898 under the heading 'Improvements in Flushing Cisterns, water-waste preventers, siphon-discharge'. A patent does not give you the right to make, use or sell an invention. Rather, a patent provides the right to exclude others from making, using or offering for sale the patented invention.

On face value Giblin's patent improvements may appear minor, when, in fact, Giblin's patent incorporates two significant improvements, one in particular is novel, quite new and original, the principle of which has lasted over 100 years. Often the best ideas are simple, uncomplicated commonsense as in the case of patent 4990 invented by a working plumber, which was not unusual. Amongst the most prolific inventors of sanitation equipment during the Victorian period were George Jennings, Frederick George Underhay, S Stevens Hellyer, John Shanks, John Bolding and Thomas Crapper, all of whom were first and foremost plumbers.

With reference to Joseph Bramah's complicated over-engineered patent valve water closet of 1778 - many distinguished engineers regarded Bramah's valve closet as the best type of water closet, a position it held for almost a century, superseded by a simple non-mechanical all ceramic wash down pan and trap incorporating an efficient flushing rim, manufactured at a fraction of the cost of a Bramah. The simple wash down closet is more efficient, hygienic, requires less maintenance and is environmentally friendly using far less water.

With further reference to Giblin's patent illustration, siphon plate 'd' has an upward inclined edge where it abuts the vertical siphon 'a' which is mirrored at the corresponding junction with chamber 'b' – this improvement ensures

Registered Grease Traps.

(No. 138,873.)

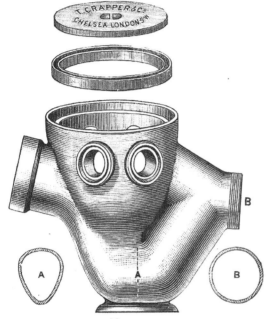

No. 540.

No. 540. Complete with Galvanized Air-tight Cover and Frame, as shewn ... 19/-

If with Grating in lieu of Cover *Less* 1/6

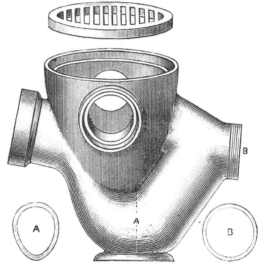

No. 541.

No. 541. Complete with Galvanized Grating, as shewn ... 17/6

If with Air-tight Cover in lieu of Grating *Extra* 1/6

Registered Flushing=rim Grease Trap.

(No. 255,985.)

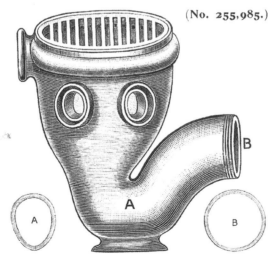

No. 542.

No. 542. Complete with Galvanized Iron Grating 26/-

If with Air-tight Cover and Frame Extra 1/6

maximum suction to activate the siphonic action whilst encouraging free passage of water when the siphon is called into use. However, the most significant aspect of Giblin's patent is the dished top of siphon chamber 'b' causing water to collect and pool thus preventing premature entry of air around rod 'c' – a simple and effective commonsense solution to a problem, which no one else had thought of before. This basic principle is replicated in modern syphons with the formation of a small raised well around the syphon rod, which when flooded stops entry of air thus preventing the premature breakdown of the desired siphonic action.

What was Albert Giblin's connection with Thomas Crapper? Giblin, at the time of his patent in 1898, was 29 and lived at 45 Barclay Road, Fulham in a shared house (flat) with his wife and young son. He was employed as a plumber. The 1911 census confirms that 42 year old Giblin progressed up the social ladder, residing at 148 Crossbrook Street, Waltham Cross, Cheshunt Middlesex. His stated occupation was water waste inspector for the local water company.

Other writers have suggested that Giblin may have worked for Thomas Crapper. This is plausible in 1898, especially given Giblin's close proximity to Crapper's Marlboro' Works, Chelsea, in fact he could easily have walked to work. Unfortunately there are no surviving employee records for T Crapper & Co for that period to refer to.

Thomas Crapper in 1898 was already famous and would have been well known to every working plumber in London. Crapper also supplied the trade with all manner of plumbing materials, pipe, fittings, lead, sanitary ware etc. Albert Giblin, without question, was acquainted with Thomas Crapper and his long established reputation. Thomas Crapper openly advertised patent 4990 as 'his' valveless waste preventer, later known as No.814. He could not possibly have made this claim without having already acquired Giblin's patent or at least been assigned a patent licence.

In normal circumstances, if the inventor is an employee and makes an invention during the course of his work, the rights belong to the employer. One possibility, considering Giblin's limited circumstances in 1898, suggests the cost of taking out the patent may have been a financial strain on his family, it is probable he could not afford to develop, manufacture, market and sell his invention without someone like Thomas Crapper. The fact is there was no one better than Thomas Crapper.

Patent
Drain Plug for Testing Drains.

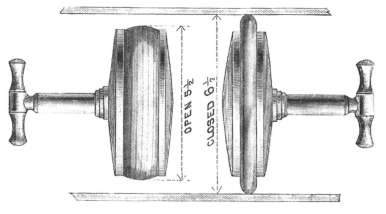

No. 392.

			4 in.	6 in.	9 in.
No. 392.	...	...	9/3	13/6	19/6

Patent
Inspection Plug and Drain Stopper.

No. 22,702.

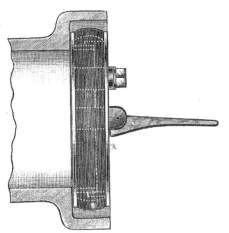

No. 393.

			4 in.	6 in.	9 in.
No. 393.	...	...	8/3	12/6	18/6

The most likely scenario, in common with many inventions, is the inventor sells his patent rights to someone who is capable of making it work commercially and turn a profit during the relatively short life of the patent. Once the patent is granted the patent holder has a limited period of time to exploit the invention whilst incurring patent renewal fees which increase as the patent gets older. Recent figures indicate the average cost of obtaining an EU patent and maintaining it for a ten year period is around €32,000.

P J Davies, a well-known Victorian plumber, teacher, inventor and author of *Practical Plumbing* (1885), claimed to be the first to make a water waste preventer upon the principle of a suction diaphragm in 1868, whilst still a journeyman plumber. Davies patented a piston diaphragm WWP which he then sold to Tylors, it later became known as the 'Waste-not Suction Closet Valve'. Davies warned potential inventors about the pitfalls and expense of taking out a patent stating *'very little is new in this world.'*

Thomas Crapper patented many of his own inventions, suggesting when Albert Giblin approached him he recognised the possibilities, seized the opportunity to acquire, develop and manufacture a non-working invention with potential. Thomas Crapper legitimately exploited and very successfully promoted CRAPPER'S VALVELESS WASTE PREVENTER PATENT 4990.

For many years Thomas Crapper promoted his Leverett cistern (named after an old Chelsea family and street) incorporating the popular fail-safe bell syphon. This was a very effective valveless water waste preventer, however, the heavy cast iron bell syphon had a habit of clanking loudly during operation, advertising the use of the WC much to the annoyance of the public, particularly when the water closet was commonly introduced into internal bathroom locations during the latter Victorian period.

Crapper employed braziers and coppersmiths in his brass foundry to manufacture bespoke copper flushing syphons to his own specification. He would supervise designs, make any modifications and apply his own rigorous performance testing.

Thomas Crapper, aware of public dissatisfaction with noisy cisterns, modified and improved Giblin's original invention so that, most importantly, it worked silently. The great success of patent 4990 must surely be attributed to Thomas Crapper as the inventor of the Silent Valveless Water Waste Preventer known as No.814. Thomas Crapper added a small bore copper syphon pipe to patent 4990, connected just below the crown of the copper

Stop Cock Boxes.

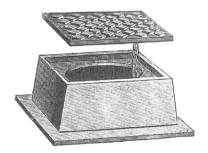

No. 470.　　　　　　No. 470a.　　　　　　No. 471.

No. 470.　11 × 8 in.　　...　　3/9　　No. 471.　5¼ × 4½ in. (Hinged)　2/2

„ 470a.　Ditto　(Hinged)　3/9

Patent Screw

(No. 11,814)

For use in connection with Manhole Covers to facilitate their removal from Frames.

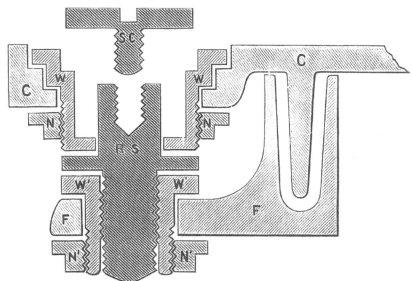

No. 472.

F　Frame to receive Cover.

C　Cover.

W　Washer to fit into Cover.

N　Nut to fasten ditto.

W¹　Washer to fit into Frame.

N¹　Nut to fasten ditto.

R.S　Raising Screw.

S.C　Screw Cap.

No. 472.　Gun Metal Screw　　　...　　6/9　each

syphon in the down flow direction. The small bore vertical leg, shown in Crapper's 1919 advertisement, absent in Giblin's patent, terminated slightly above the base level of the syphon suction chamber. This small bore pipe worked as a combined silencer and mini injector, effective simultaneously when the siphonic flush was activated. The obvious benefit of Crapper's improvement was to quickly introduce air just as the cistern emptied, equalising the air pressure within the syphon chamber, thus reducing noise upon completion of the flushing cycle. In normal circumstances air re-entering a syphon chamber at the base breaks the siphonic action, however in doing so causes an audible gurgling noise which echoes loudly within the empty cistern.

Crapper further modified patent 4990 so that it operated efficiently with a modern style flat bottom cistern box, as opposed to the old well bottom style of cistern shown in Giblin's patent. In the Author's opinion Giblin's original patent combined with Crapper's modifications and improvements made a substantial contribution towards the high performance water closet flushing syphon cistern still in use today. Crapper's ingenious arrangement was novel and allowed him to advertise to the general public CRAPPER'S SILENT VALVELESS WATER WASTE PREVENTER available in five flat bottom cistern box finishes – painted cast iron, galvanized, porcelain enamelled, earthenware and lead lined cases.

Albert Giblin in his patent specification stated *'an effective discharge can be obtained when the cistern is in any state between about half full and full.'* Following Thomas Crapper's own rigorous flushing performance testing on patent 4990 he substituted Giblin's ambiguous, potentially misleading claim for his own confident, bold statement *'will flush when two-thirds full.'*

Albert Giblin clearly had a very inventive mind, at least as far as water closet flushing cisterns were concerned, for he went on to patent two further improvements in flushing cisterns. In 1912 he was granted Patent No.10213, apparatus for actuating syphon bells which became known as The Roller Syphon Cistern. In 1920 he was granted Patent No.167433 which related to apparatus for bell type cisterns in which the sump for the bell was provided with a loose flanged collar, this became known as the Easy Flush Syphon Cistern. Giblin achieved moderate success with these latter two patents however without Thomas Crapper's involvement they failed to achieve the

Mica Air=inlet Valves.

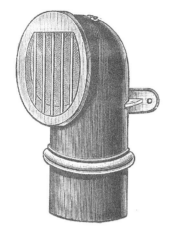

No. 431.

No. 432.

No. 433.

			3 in.	3½ in.	4 in.
No. 431.	With Brass Front, Spigot end	...	3/11	4/2	4/6
,, 432.	Ditto Socket end	...	3/11	4/2	4/6
,, 433.	Ditto Open Balance	...	4/2	4/6	4/10

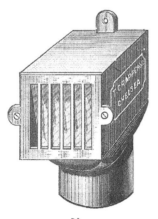

No. 434.

No. 435.

No. 436.

				3 in.	3½ in.	4 in.
No. 434.	With Brass Front, Spigot or Socket end	...		3/11	4/2	4/6
,, 435.	Ditto ditto Double faced			5/11	6/9	7/9

				7 × 5 in.	9 × 7 in.
,, 436.	Galvanized Iron	... ...	...	5/6	7/9
,, 437.	Ditto Brass Front	...	...	6/9	9/6

success of Patent No.4990 (814). Albert Giblin's descendants remain involved in the bathroom industry in 2014. Crapper's Silent Valveless Water Waste Preventer enjoyed a long life, by the 1930s it was known as the Marlboro Cistern. Crapper & Co made their patent 4990 syphon in copper, gunmetal and cast iron. Eventually lime-scale affected its performance, choking up the silencing pipe. Cheaper syphons made of ceramic were introduced in the 1930s and by 1960 the plastic syphon gained popularity on price. Leather, rubber and later plastic flexible syphon diaphragm washers replaced Crapper's syphon copper suction plate, however the basic principle remains.

In 2001, as a result of pressure from the EU, inept British politicians permitted the use of cheap cistern flushing valves, a backward step, which sooner or later leak and waste water. There is an old saying *'All cheap plumbing is bad and all good plumbing must be paid for.'*

A major contribution to sanitary engineering was Crapper's Patent Disconnecting Drain Trap 1888 (registered design 1881) referred to as an Intercepting Trap, these names are synonymous. In Thomas Crapper's day, old below ground sewerage drainage systems with a slow flow and poor ventilation were commonplace problems. The disconnecting trap was incorporated into a ventilated chamber in the underground drainage system at the boundary of the property in order to prevent foul air from the main sewage system entering the property. These obnoxious gases were poisonous and explosive and greatly feared, particularly with the advent of indoor sanitation, a luxury which also carried potential dangers for properties and inhabitants. Crapper's patent incorporated an easy access valve for clearing blockages, particularly useful in deep manhole situations.

Newly developed remote operated sanitary appliances had many applications, they were invaluable for health and hygiene in the prevention of spreading germs and also very useful for some less able bodied people. Crapper's Patent Foot Treadle 1893 was installed to the cloakroom facilities adjacent the billiard room at Sandringham House.

Crapper's Patent Seat Action Automatic Flushing Arrangement 1891 was adapted for hotels and public buildings, it was commended by the press as *'a very excellent arrangement – simple and effective.'* This patent is sometimes referred to as the 'bottom slapper'.

Crapper's patent Inspection Plug and Drain Stopper of 1894 with fast action lever operation is still in use today.

Closet Fittings.

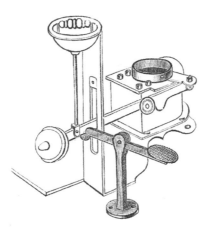

Patent Closet Treadle

(No. 11,604)

Mechanical Operator for opening
Valve with foot when apparatus
is used for slops.

Can be fitted to existing

Closets.

———

No. 252. ...   7/6

No. 252.

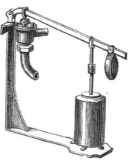

No. 253. $\frac{3}{4}$in. Stool Valve, with
Union, and Zinc Regu-
lator on half-frame ... 16/9

,, 254. Brass Sunk Dish with
China or Ebony Handle 7/3

No. 255. Stool Valve, with Union
$\frac{3}{4}$ 1 $1\frac{1}{4}$ $1\frac{1}{2}$in.
6/8 8/9 12/6 16/9

,, 256. Bellows Regulator with Tap
Zinc 5/6 ; Copper 7/3

No. 253.

No. 257.

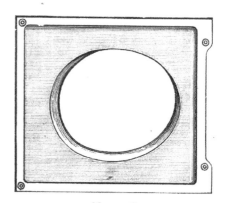

No. 258.

No. 257. Earthenware Slop Top and Dish in one piece, White or Ivory   ...   £1 7 6

,, 258 Ditto   ditto   White or Ivory 0 10 6

Crapper's Improved Bath Waste and Overflow 1889 incorporated an anti syphonage pipe connection to the trap, designed to prevent loss of trap water seal and the entry of foul gases from the sewer. Crapper's trap design minimised waste water syphonage noise as the bath emptied. Modified and today known as an Anti-Vac Trap.

Crapper's Patent Screw Mechanism 1893 was a novel solution to the common problem of raising heavy cast iron manhole covers, particularly useful if maintenance was neglected.

Crapper's improved ornamental syphon cistern 1890 was a leap forward in sanitary ware design incorporating concealed cistern fixings and provision for connecting concealed pipework. This innovation was useful for both interior design projects and institutions where vandalism was a problem.

Crapper's registered flush down water closet of 1890, with flushing rim and improved trap, uniquely incorporated The Prince of Wales Feathers emblem. It was a throne fit for a future King!

The following patents / designs have historically been attributed to Thomas Crapper, however currently remain subject to further verification and in consequence are not included, at this time, in the patent and registered design schedule hereafter.

➢ Crapper's Patent Automatic Siphonic Flushing Tank
➢ Crapper's Patent Trough Closet
➢ Crapper's Improved Reverse Action Ball Valve
➢ Crapper's Cantilever Water Closet
➢ Crapper's Patent Flusherette Valve

The "Cecil" Porcelain Slop Sink,

With Flushing Rim.

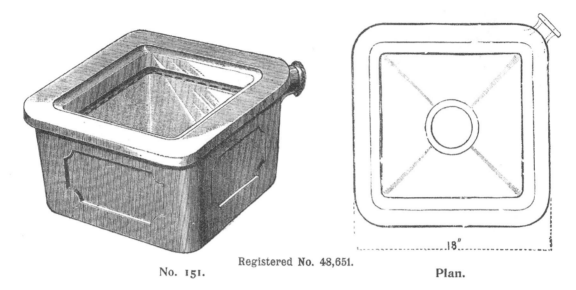

No. 151.   Registered No. 48,651.   Plan.

No. 151. "Cecil" Slop Sink, White Glazed

13 × 13 in. inside   ...   £2 7 6

If Glazed outside   Extra 12/6

,, 152.   White Glazed

22 × 22 in. outside   ...   4 7 6

Brass Outlet and Union   Extra 14/6

The Flushing Rim is so arranged that the whole of the interior is cleansed at each flush.

This Sink was awarded the Medal of the Sanitary Institute, at Worcester, 1889, and again by the Sanitary Institute at Brighton, 1890, being the highest possible award in both cases.

Can be made to special dimensions if required.

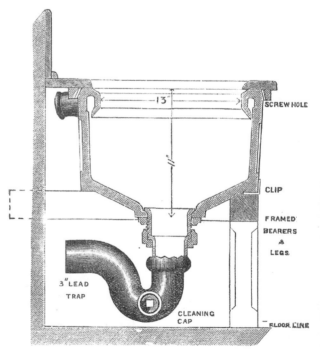

Section shewing method of fixing Wood Rim, and suggestion for fitting complete, with Trap.

T Crapper & Co Patents

Patent No.	Year	Description
1628	1881	Air inlet for ventilating houses and drains.
10332	1888	Crapper's disconnecting drain trap.
3964	1891	Crapper's patent seat action automatic flush down WC arrangement.
11604	1893	Improved means for operating the mechanism of water closets via a foot treadle.
11814	1893	Inspection chamber rising screw mechanism.
22702	1894	An improved expanding plug for drain pipes and traps.
4333	1896	Improved pipe joint, applicable for connecting pipes and traps, closets and traps and the like.
4990	1898	Improvement in flushing cisterns, water waste preventer, siphon discharge. Albert Giblin original patentee, developed, modified and improved by Thomas Crapper.
6029	1903	Improvement to stair rods.

Combined Bath Sets.

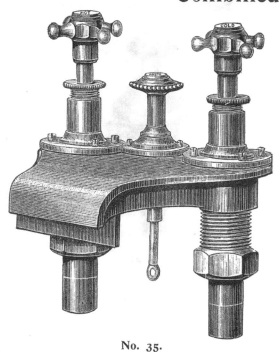

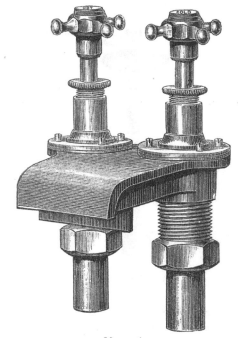

No. 35.   **No. 36.**

No. 35. Yellow Metal ...   ...   £1 2   No. 36. Yellow Metal £0 18 3

       If Plated, Extra 9/6   If Plated, Extra 7/6

Improved Bath Waste and Overflow

(Registered No. 120,114.)

Bath Washer and Grate (or Plug)

With Fly=Nut and Union.

No. 38.

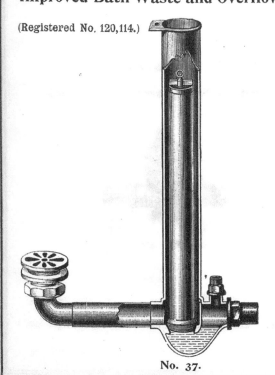

No. 37. Strong Zinc Stand Pipe with Crapper's Registered Trap (Brass), Union to connect Waste Pipe, Union to connect to Anti-Syphonage Pipe, and Brass Washer and Grating, complete as shewn   £1 9 9

If fitted with Copper or Brass Stand Pipe   ...   Extra 7/-

Inches 1½   2

„ 38. Bath Washer and Grate (or Plug) Fly-Nut and Union 8/-   11/-

No. 37.

T Crapper & Co Registered Designs and Trade Marks

Design No.	Year	Description
81187	1881	Trade Mark for Thomas Crapper & Co.
105149	1881	Crapper's disconnecting trap.
120114	1889	Crapper's Improved Bath Waste and Overflow.
138873	1890	Grease trap, Crapper & Co.
145823	1890	Crapper's improved registered ornamental flush down water closet. This design included The Prince of Wales feathers emblem displayed on the closet pan.
149284	1890	Crapper's improved ornamental syphon cistern.
231040	1894	The improved Kenon disconnecting trap, Thomas Crapper & Co.
245045	1894	White glazed blocks for inspection chamber floor.
255985	1894	Flushing rim grease trap.

➢ Patent No.724 1897 was granted to George Crapper and Robert Marr Wharam for improvements relating to Automatic Syphon Flushing Tanks.

➢ Patent Law: protects the way things work and what they are made of, it lasts up to 20 years.

➢ Registered Design: protects how something looks, lasts 5-25 years.

➢ Trade Mark: protects a trading brand and image of a company – lasts forever.

Syphonic Latrine.

This system of Latrine forms the best possible Closet Range for Schools, Workhouses, Factories, &c. All the objections hitherto raised to Trough Closets are removed.

In this system all the water from Cistern is passed through the Arms of Basins, thereby washing the whole surface of Basin, and preventing any part getting fouled. The contents of Basins and Main Discharge Pipe are removed at each flush by Syphonic Action; the Main Pipe is refilled, and a large surface of clean water left in Basins. Basins of Strong Fire Clay, White Enamelled inside.

No. 208. Complete with Galvanized Cistern and Flush Pipe, for Six Persons **£22 12 6**

Quotations given for Ranges for any number of Persons.

CHAPTER

HARD TIMES

Henry and Thomas Crapper were brothers, born within a few years of each other at Thorne Quay, Yorkshire. Both relocated to London where they lived and worked as plumbers just a few miles apart. Henry and Thomas Crapper both died in the month of January from cancer age 72 and 73 years respectively, both brothers survived their wives.

One brother achieved fame and fortune, his celebrated life commemorated on a white marble memorial. The other died a destitute pauper in the workhouse infirmary, laid to rest in an unmarked common grave. Few people are destined for greatness whilst many often struggle to survive. Thomas and Henry Crapper lived in very different circumstances. Thomas in Battersea and Penge, Henry in Lambeth and Camberwell, in reality their lives were worlds apart.

Henry Crapper, born in 1833, apprenticed age 14 to Thomas Cundey, Master Plumber employing three men at Bridge Street, Rotherham, Yorkshire. Henry's apprenticeship fee was paid for by his father. Henry lived above the shop with the Cundey family and another apprentice. Cundey was an agent for Kirkwood's registered water closets, awarded a prize medal when exhibited at the Great Exhibition in 1851. Once qualified Henry worked as a journeyman plumber in Rotherham, then in nearby Sheffield where he established his plumbing business c1858. Henry appears to have been doing well, in 1860 he married local girl Sarah Walker at Sheffield. Their oldest sons Thomas Crapper born 1864 followed by Henry junior in 1866 were born in Sheffield.

Misfortune struck Henry in 1867 when his business failed. A notice appeared in the London Gazette. *'Henry Crapper of Sheffield, Yorkshire, plumber and gas fitter having been adjudged bankrupt 4th April 1867.'* What caused Henry's financial difficulties is not known, perhaps illness, bad debts or some carelessness of drink or gambling, we do not know. Henry decided to relocate to London where his brothers George, Thomas and William were all doing well as

Cane Glazed Kitchen Sinks.

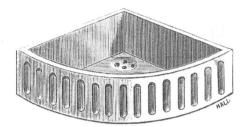

No. 143.

Across Corners.	Back to Front.	Length of Sides.	Depth.			Price.
16 in.	16 in.	11½ in.	5 in.	...	...	3/9
20	20	14	5	...	...	4/8
24	20	17	5	...	...	6/3
26	20	18½	5	...	...	7/10
28	22	20	5	...	...	10/3
34	24	24	5½	...	...	13/9

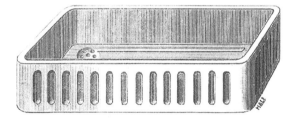

No. 144.

20 × 15 × 5 in.	...	6/3	33 × 18 × 5 in.	...	12/6
22 × 16 × 5	...	7/6	32 × 20 × 6	...	16/-
24 × 18 × 5	...	9/3	36 × 22 × 6	...	20/-
27 × 18 × 5	...	10/3	42 × 24 × 6½	...	27/6
30 × 18 × 5	...	11/6	48 × 24 × 6½	...	32/6

These Sinks also supplied White inside, for which add 50 per cent.

plumbers. Henry was not alone in his bankruptcy, in 1868 Thomas Crapper also suffered a financial jolt when he became a main creditor in the bankruptcy of builder George Godbolt of 289 King's Road, Chelsea. Thomas Crapper was appointed trustee and successfully negotiated the assignment of Godbolt's estate, securing payment to creditors by instalment amounting to 10 shillings in the pound. (London Gazette May 1868) Thomas Crapper established his business in 1861, by 1868 he was employing 24 people.

Henry found work in London as a journeyman plumber, he rented a small house in Edmund Street, Lambeth, Camberwell. It is uncertain what contact Henry had with his industrious younger brother Thomas or older brother George who accommodated their parents at his Chelsea and Acton homes, both parents passed away at Acton during October 1873. Henry's situation improved and he relocated to Surrey Road, Peckham where his children Emily, George, James, Bryan and Lucy were born.

The census for 1881 enumerates: Henry Crapper 47 journeyman plumber, sons Thomas 17 and Henry junior 15 both apprentice plumbers, following in their father's trade. Tragedy struck in January 1886 when Henry's wife Sarah died age 50, leaving him to provide and care for Lucy 6, Bryan 8, James 10, George 12 and Emily 15. Henry's circumstances deteriorated, unable to cope his family was separated. Lucy became a resident at Halstead Industrial School for girls in Essex. Bryan was sent to Somersetshire Industrial School for Boys at Bath, James was an inmate aboard The Shaftesbury a ship for 500 boys moored at Grays on the River Thames, utilised as an industrial school training ship. George was more fortunate, he was dispatched to Rotherham in the care of his mother's family.

Henry never recovered from the loss of his wife and family, he resided in one room at Hollington Street, Camberwell with his daughter Emily who was employed as a mantle maker. The final years of Henry Crapper's life were, sadly, as an inmate pauper / retired plumber at Constance Road Workhouse, Camberwell.

Included in a report of The Chelsea Workhouse Committee, 13th October 1866, *'Thomas Crapper, plumber, be employed to remove the bath from the old receiving room on the female side of the workhouse, to the present receiving room.'* Chelsea Workhouse housed: 24 men, 176 women, 64 boys, 76 girls, total 340. Camberwell Workhouse housed 850-1,000 inmates.

IN
LOVING MEMORY OF
THOMAS CRAPPER.
WHO DIED JANUARY 27TH 1910.
AGED 73 YEARS.

Thomas Crapper 1836-1910
Beckenham Cemetery formerly Elmers End

Photograph courtesy of Aysen Slack © 1991

Camberwell Workhouse records confirm that Henry Crapper was resident in 1901 at Constance Road. Between August and December 1904 he was admitted and discharged on 30 separate occasions. The description of Henry when admitted often recorded him as *'widower, plumber, Camberwell, destitute, aged and infirm.'* On Saturday 8th October 1904 Henry was recorded on admission as being of *'unsound mind.'* On the 31st December 1904 he is described as *'aged, infirm and fits – discharged at own request.'* Henry Crapper died 24th January 1905 at Constance Road Infirmary, Camberwell aged 72 years, his daughter Emily was present at his death.

Henry's children all survived, his eldest son Thomas continued working as a plumber in Camberwell, he also had a son named Thomas Crapper. Henry Crapper junior continued to work as a plumber in Southwark, he married twice but had no children. George became a successful roofing and building contractor at Rotherham employing brothers James and Bryan. Lucy went into domestic service as a housemaid to Mr Dickenson a portrait painter of Marylebone, London. Emily married Robert Groves in 1897 at Newington, London. A photograph of a rotund, jolly looking Emily age 50 residing in Camberwell has been posted on a family history website by her descendants. Emily, born in 1871, experienced tragic 'hard times' in her early life, hopefully she was proud of her famous, if distant, uncle Thomas Crapper, Sanitary Engineer 'By Appointment to his Majesty the King'.

Thomas Crapper died 27th January 1910, he was interred at Elmers End Cemetery, Anerley, Kent, reunited with his beloved wife Maria. Thomas Crapper left an estate equivalent to £1.5m in 2014. His executor was his nephew Charles Crapper, eldest son of his brother George and executrix Emma Crapper, niece and seventh child of brother George Crapper. Thomas left cash bequests to his wife's sister, nephew and niece. He left his house 12 Thornsett Road, Anerley, Kent together with all his household furniture, linen, plate, books, pictures, prints, wines and spirits and jewellery to his spinster nieces Emma and Maria Crapper plus an annuity each of £100 per annum. Thomas bequeathed an annuity to his brother William Green Crapper of £1 per week for life. The residue of his estate was divided equally between his brother George's seven surviving children.

Emma born 1869, Maria born 1874 and their older brother Thomas Crapper born 1862 never married. In 1901 all three shared a house at 36 Prebend Gardens, Chiswick. In 1902, following the death of Thomas's wife Maria all

Improved Self=rising Closet Seat.

Specially adapted for Closet used also as Slop Receiver.

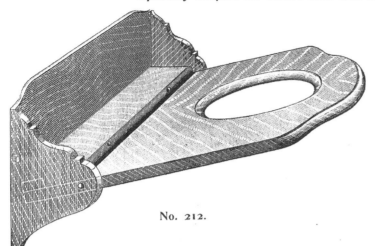

No. 212.

Mahogany or Walnut Polished ...	...	29/6
Stained Deal	...	22/6
Plain Deal	...	19/6

No. 212.

Wood Seats for Pedestal Closets.

(Second Quality.)

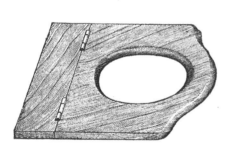

No. 213.

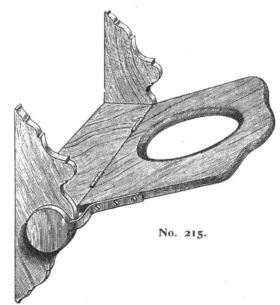

No. 215.

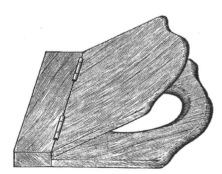

No. 214.

	Plain Deal	Stained Deal	Mahogany
No. 213.	6/9	7/9	9/9
,, 214.	12/9	14/3	16/6
,, 215.	14/9	17/3	19/6

three resided with Thomas at his Thornsett Road home. In 1911 Thomas Crapper, born 1862, was described as an invalid, he died at Thornsett Road on the 10th March 1918. He is buried at Elmers End Cemetery in close proximity to his famous uncle Thomas Crapper. Maria Crapper of 12 Thornsett Road, Anerley, Kent died 29th September 1944 at her sister's home at Great Malvern, Worcester. Emma Crapper died on the 12th August 1945 at Norcroft, Alexander Road, Great Malvern, Worcester.

Edmund Sharpe 1810-1894

Photograph courtesy of Julian Whitaker

1. Sharpe's Patent Flushing Rim
with Trap 1855

2. The Household Closet c1865

3. The Burton Closet with Sharpe's Patent
Flushing Rim c1875

4. The Cedric 1895

Evolution of Sharpe's Wash Down Water Closet

CHAPTER

TRADING PARTNERS

Thomas Crapper and Sharpe Bros

Sharpe Brothers & Co of Swadlincote, Derbyshire was a long established and major supplier of sanitary ware to Thomas Crapper & Co of London. This valued business relationship was of mutual benefit to both companies. Each prided themselves on quality, reliability, service and innovation, their trading partnership existed for almost 100 years.

Thomas Sharpe established his pottery in the small hamlet of Swadlincote in 1821 making stone bottles, Toby jugs and other wares. Thomas laid the foundation on which Sharpe's Pottery flourished for 147 years, manufacturing and exporting their products all over the world.

In 1838 Sharpe Brothers & Co produced ordinary domestic items e.g. pie dishes, teapots and chamber pots. They were awarded an honorary mention for the quality of their wares at The Great Exhibition held at Crystal Palace, London 1851. However it was a Sharpe invention and patent in sanitary ware design in 1855 that resulted in Sharpe's Pottery becoming the world's leading manufacturer of water closets, a position they held for a quarter of a century. Sharpe's Pottery is unique and significant for surviving almost 200 years on the same site and is now believed to be the oldest surviving sanitary pottery works in the world.

Sharpe Bros & Co was a key supplier of sanitary ware to overseas markets including America, Australia, New Zealand, South Africa, Jamaica, Egypt, Poland, Chile, Brazil, Peru, Canada, Ireland, Romania, Germany, Belgium, Russia, Italy, Holland, Spain, Portugal, India, Bulgaria, Sweden, Denmark and Greece.

In 1845 Edmund Sharpe visited America aboard The Cambria a wooden built paddle steam ship, sailing from Liverpool to Halifax and Boston. During

Pedestal Wash=down Closets.

"The C.C. Primrose"
(With S or P Trap).

No. 176.

"The C.C. Ivanhoe"
(With S or P Trap).

No. 177.

No. 176.				
White or Ivory	£1	9	0
White or Ivory and Printed	1	12	6
Decorated	1	15	6
Decorated and Printed ...	1	19	0

No. 177.				
White or Ivory	£1	6	0
White or Ivory, and Printed	1	9	6
Ditto, Printed inside and out	1	12	6
Decorated in Colours ...	1	16	6

"The Cedric"
(With S or P Trap).

No. 178.

"The Straightback"
(With P Trap only).

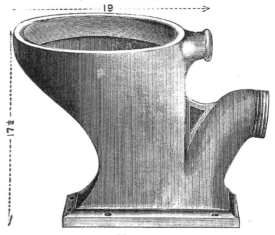

No. 179.

No. 178.				
Cane and White ...	£0	19	6
Cane and Printed ...	1	2	6
White or Ivory	1	2	6
Ditto and Printed...	1	6	0
Ditto Printed inside and out ...	1	7	6

No. 179.				
Cane and White ...	£1	3	6
Cane and Printed ...	1	7	6
White or Ivory	1	7	6
White and Printed inside	1	11	6
White and Printed inside and out	1	14	0

a visit to the House of Representatives at Washington Edmund noticed pine boxes with grass in the base being used as spittoons. Edmund recognised an opportunity for improvement and upon his return to England he registered a design for a large pottery spittoon known as the Congressional Spittoon. Edmund's visit strengthened demand for Sharpe's products.

During Edmund Sharpe's return trip from America he met Frederick Douglas a former escaped slave. Douglas was a controversial figure and reportedly caused a brawl on the deck of The Cambria between pro and anti slavery passengers. During the return trip Edmund also befriended the Hutchinsons, an American group of three brothers and their sister who were famous professional singers visiting and touring Great Britain for the first time. Edmund accommodated the singers at his home in Swadlincote, Derbyshire. In return the Hutchinsons put on an exclusive concert perfor-mance at Sharpe's Pottery works for Edmund's 165 pottery workers. When it was time for the singers to return to America Edmund Sharpe presented each of them with a tea chest containing a full dinner service, made in his own pottery, as his personal gesture of Anglo-American friendship.

The Sharpe family ran the pottery for 102 years, the Whitaker family followed and still own the freehold of the old pottery. Sharpe's Pottery, Swadlincote serviced most of the leading sanitary ware suppliers with their products or badged for specific customers e.g. Thomas Crapper & Co, George Jennings, Dent & Hellyer, John Bolding, Finch & Co, Baxendale & Co, Nicholls & Clarke etc. Sharpe's also manufactured sanitary ware for other potteries e.g. Edward Johns at Armitage, Doulton & Co, Shanks, Thomas Wragg, Morrison, Ingram & Co etc.

Edmund Sharpe's historically important 1855 Patent Box Rim Flushing Water Closet invention was ingenious and of international significance. Edmund solved the problem that had confounded every water closet inventor since Alexander Cummings' patent of 1775. Sharpe's patent flushing rim water closet used less water and was more efficient in flushing and cleansing than any other water closet ever invented. Sharpe's patent specification contained the words 'wash down-ward' and Sharpe's later referred to this as 'The Household Wash-down Closet Pan' which featured in many sanitary ware catalogues including Thomas Crapper & Co. Sharpe's patent flushing rim pans were commended by many leading sanitarians including George Jennings, P J Davies and S Stevens Hellyer.

Shampooing Apparatus.

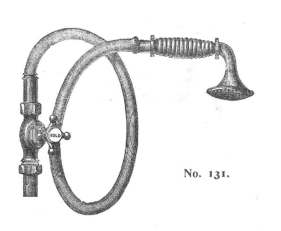

No. 131.

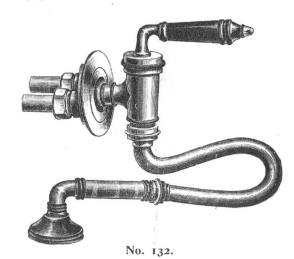

No. 132.

		Yellow Metal.	Plated.
No. 131. Valve, Brass Rose, Horn Handle and India Rubber Tube ...	½-in.	21/-	25/6
	¾-in.	28/-	33/6
„ 132. Valve, Brass Rose and India Rubber Tube		34/6	42/-

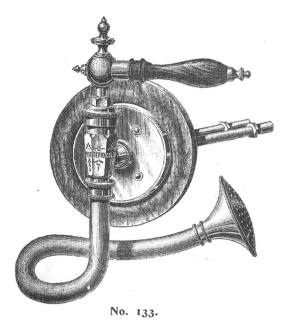

No. 133.

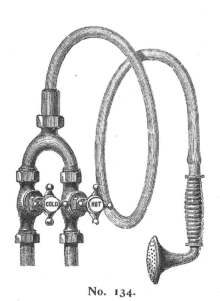

No. 134.

		Yellow Metal.	Plated.
No. 133. Valve, Engraved Hot, Cold and Tepid, mounted on Polished Block with Brass Rose and India Rubber Tube		45/-	54/6
„ 134. Valves, Brass Rose, Horn Handle and India Rubber Tube ...	½-in.	32/-	39/6
	¾-in.	46/-	55/6

P J Davies, in his book *Practical Plumbing* (1885) wrote *'Sharpes Patent Flush Rim Closet is a decided improvement and which is now universally used.'* The Parkes National Museum of Sanitation contained exhibits of old and new sanitary ware designs including a number of water closets made by Sharpe Bros & Co. The Household Water Closet was noted in the Museum schedule 1891 *'... this closet had been fixed in 1883 and was in daily use, every part is open to inspection, probably the first in England arranged in this way.'*

Sharpe Bros manufactured Jennings' Monkey closets and made the world's first one-piece earthenware valve closet for Jennings in 1862. Thomas Twyford c1870 was advertising closet basins incorporating Sharpe's patent flushing rim, advocating that Sharpe's box rim flush was the best on the market. Edmund Sharpe invented the 'box rim' which is synonymous with the 'wash down' water closet. The Bathroom Manufacturers' Association in their Bathroom Academy information sheet 2013 stated *'The wash down – box rim is particularly suitable for domestic houses and hotels. The flush is better controlled, creates least turbulence, less aerosol effect and is the quietest flush.'* Edmund Sharpe's box rim invention remains the gold standard for all wash down water closets.

Sharpe's Pottery's archive includes workers employment contracts from 1860, catalogues, minute book and records of Sharpe's involvement with several Pottery Manufacturers' Associations. These records contain numerous references to Thomas Crapper & Co. Sharpe's supplied Crapper with many patent water closets, some of which are included in Thomas Crapper & Co's 1895 catalogue e.g. Capstan, Cedric, Primrose, Ivanhoe, The Household, Marlboro, Straightback and Swift as well as many valve closet pans, wash basins, urinals etc.

The Plumbers' Journal of 1935 noted *'Crapper's still do a considerable trade in valve closets for use by many old English gentry who will have no other type.'* The demand must have remained for this product as Sharpe's continued to supply these to Crappers ex-stock. T Crapper & Co, brass founders and Sharpe Bros, potters were the last two companies, in partnership, to manufacture valve water closets as part of their standard product range. Sharpe Bros sales ledger 1947-51 includes orders delivered to T Crapper & Co Ltd, 120 King's Road, Chelsea. During this period T Crapper & Co purchased 300 Sharlin closets, Sharcote closets, Shark closets, Marlboro

Cast Lead Traps.

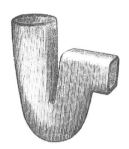

No. 394.

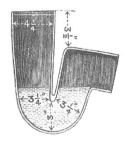

Section.

No. 394.   4 in. Anti-D Trap for W.C.'s 7/-

    3½ in. to 2 in   ditto,   for Sinks, Baths, etc.   ...   4/-

    1¼ in.   ditto,   for Lavatories, etc...   ...   2/-

    1¼ in.   ditto,   with 2½ in. mouth   ...   2/3

3½ in. and 1¼ in. Traps can be had with Screw Cap ...   Extra   1/-

No. 395.

No. 395.   Pedestal Closet Trap,   P 8/3 ;   S 10/9

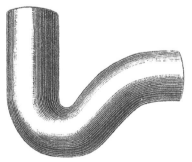

No. 396.

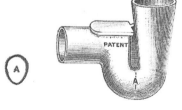

No. 397.

No. 396.   4 in. P Trap   ...   6/3   No. 397.   4 in. Closet Trap ...   6/6

closets and 37 valve closet basins (Bramah type), the latter were for both the new and replacement market.

Sharpe's Pottery ceased sanitary ware production in 1968, however following extensive restoration it reopened in 2003 as Sharpe's Pottery Museum & Community Resource Centre. Sharpe's Pottery's listed original bottle kiln and substantial attached workshop buildings are an evocative reminder of the historic South Derbyshire pottery industry, which was once second in size only to Stoke-on-Trent.

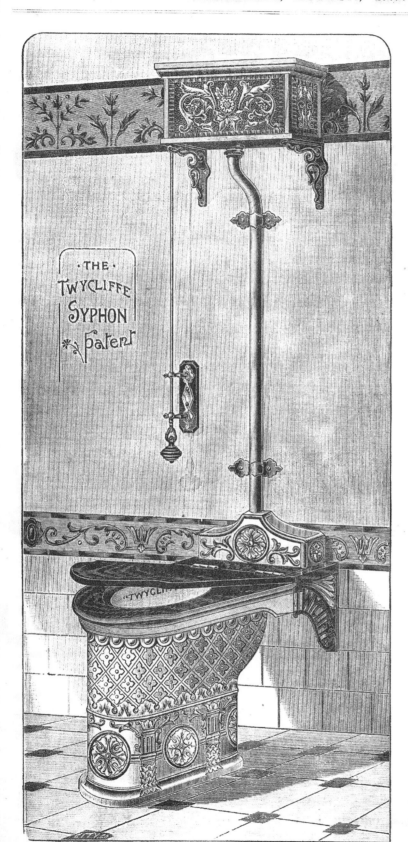

The "Twycliffe"
PATENT
Syphon Closet Basin.

This Improved Syphon Closet is as simple in Construction as an ordinary "Wash-down," is without Mechanism, and requires only an ordinary Syphon Cistern with One Flush Pipe. Its action is quick and noiseless.

The Depth of Water in Basin is 7 in.

The Surface of Water is 12 × 10 in.

The Water Seal in Trap, which is $3\frac{1}{2}$ in., forms an effectual barrier to the passage of sewer gas, and cannot be broken by Syphonic Action, or the emptying of slops in Basin, &c.

No special directions are required as to fixing, except that it must not be fixed over a Trap, as no Second Trap is required.

The Connection between the Outlet of Trap of Basin and the Lead Soil Pipe is made by Patent Porcelain-Metal Joint, which being a Wiped Joint is a most perfect and permanent Air-tight Connection.

Made in Porcelain—Plain, Decorated, and with Raised Ornamentation; and Strong Fire Clay—Porcelain Enamelled.

Made with Trap Outlet to suit requirements of the London County Council— and also with Outlet in centre of Basin.

———

No. 159. Complete, including Porcelain Basin with Slop Top and Trap in one piece, White or Ivory, "Corinthian" Pattern in Relief, with Paper Box to match; Brass Connection to Supply Pipe; Patent Porcelain-Metal Wiped Joint to Outlet and Double Lead Bend (unattached); Mahogany Seat with Flap, Best Quality, with Galvanized Brackets; Mahogany Cistern, Copper Lined; Brass Brackets; Brass Flush Pipe, 6ft., and Clips; Pendant Pull with Block

£12 6 6

With Seat without Flap, Cast Iron Galvanized Cistern and Brackets, Galvanized Cast Iron Flush Pipe and Chain Pull and Handle

£9 8 6

Thomas Crapper & Thomas Twyford

Thomas Crapper and Thomas Twyford are both iconic names in the British sanitary ware industry. It's not a coincidence that the sanitary ware section of Thomas Crapper & Co's 1895 catalogue was dominated by Messrs Twyford and Sharpe's products, as they were the leading and oldest established sanitary ware potteries of the period. Other regular suppliers were Joseph Cliff & Sons, James Woodward & Rowley, Geo Howson and Doulton & Co.

Crapper's 1895 catalogue included Twyford's very latest products manufactured in Cliffe Vale porcelain enamelled fire clay. Prominently featured were The Adamant urinal range, lavatory basin range, wash tub range and syphonic latrine range. Twyford specifically developed these exclusive products for the booming development of clubs, hotels, railway stations, underground urinals, schools, workhouses and factories. Also featured were Twyford's water closets: The Twycliffe, Deluge, Cardinal and Zone.

Thomas Twyford, born 1827, together with his brother Christopher, born 1826, formed a partnership sometime in the early 1850s as earthenware manufacturers of Hanley, Staffordshire. The brothers were born into a pottery family, their father William Twyford, 1795-1881, worked as a pottery printer and Christopher initially followed his father's trade, whilst Thomas was, in 1841, recorded as a pottery turner's apprentice. Christopher Twyford, in the 1851 census, is enumerated as a potter's mould maker and his wife Ann a pottery paintress, whilst Thomas, 24, was employed as a china burner residing at Blockhouse, Worcester with his wife Sarah, a shopkeeper, and their son Thomas (William) Twyford age 1.

A notice appeared in the London Gazette on the 15th July 1859 announcing *'The partnership between Thomas Twyford and Christopher Twyford carrying on the trade as earthenware manufacturers in High Street and New Street, Hanley in the style of T and C Twyford was, on the 9th July, by mutual consent dissolved. All debts owing to the partnership to be paid to Thomas Twyford who will continue the manufactory in Bath Street. The manufactory in New Street will be continued by Christopher Twyford.'* Christopher Twyford died in 1869 and left estate under £200.

Thomas Twyford manufactured ceramic beer pump handles and spirit barrels, whilst establishing a growing trade and reputation as a specialist sanitary ware manufacturer. Although Thomas held no patent of his own, his business prospered making basic closet pans, traps, valve and pan closet

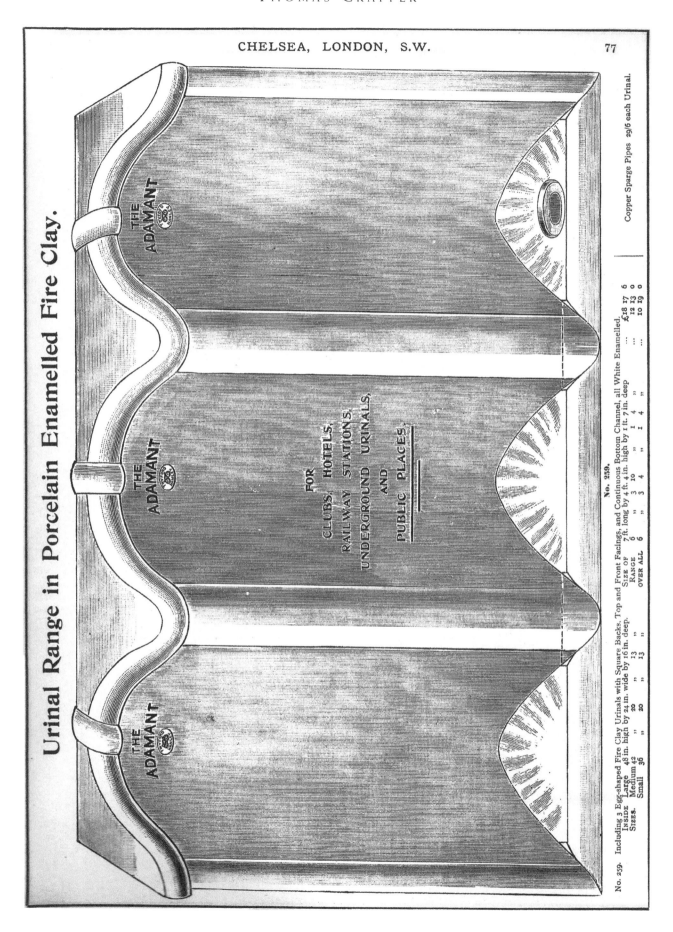

Urinal Range in Porcelain Enamelled Fire Clay.

THE ADAMANT

THE ADAMANT

THE ADAMANT

FOR
CLUBS, HOTELS,
RAILWAY STATIONS,
UNDERGROUND URINALS,
AND
PUBLIC PLACES.

No. 259.

No. 259. Including 3 Egg-shaped Fire Clay Urinals with Square Backs, Top and Front Facings, and Continuous Bottom Channel, all White Enamelled.
Size of 7 ft. long by 4 ft. 4 in. high by 1 ft. 7 in. deep
Inside Large 48 in. high by 24 in. wide by 16 in. deep.
Sizes. Medium 42 „ 20 „ 13 „
Small 36 „ 20 „ 13 „

	RANGE	6 „ 3 10 „ 3 4	... £18 17 6
	OVER ALL 6 „ 3 4 „ 1 4 „	... 12 13 0
		... 10 19 0

Copper Sparge Pipes 29/6 each Urinal.

basins, urinals etc. When Thomas Twyford died in 1872, the pottery passed to his son Thomas William Twyford, 23. With an infusion of youth, ambition and inventiveness Twyford's business took off c1878 resulting in Twyford's pottery producing their own market leading products e.g. The National, Unitas, Deluge and Twycliffe.

Between 1884 and 1892 T W Twyford was granted thirteen patents for sanitary ware. By the 1880s Twyfords were the largest sanitary ware manufacturer in the UK, if not the world. Twyford's exhibited at every sanitary and health exhibition, winning many medals and awards. In 1887 the company built a new factory at Cliffe Vale costing £50,000 by which time they had a workforce of 400.

The sanitary ware manufacturing industry boomed between 1875-1900 with British sanitary potters Twyford, Sharpe, Doulton, Johns, Johnson Bros and many more supplying the world with quality products. Twyfords continued to expand, they started making fireclay sanitary ware in 1890. In 1901 they built a new factory in Germany and in 1911 opened a new fireclay plant at Cliffe Vale.

Thomas William Twyford valued long trading partnerships, he supplied ware to Thomas Crapper & Co until he died in 1921. Thomas W Twyford amassed a great fortune amounting to £236,000. Twyfords became a public limited company in 1919 which resulted in the founding family losing control of the business. In 1950 Twyford's invested £400,000 installing gas fired kilns at Cliffe Vale and the latest tunnel kilns at Alsagar.

In 1971 Twyfords were taken over by Reed International Ltd then acquired by Caradon Ltd in 1985. In 1999 Twyford Bathrooms were granted their first Royal Warrant by Appointment to Her Majesty Queen Elizabeth II. HSBC sold Twyford Bathrooms in 2001 to the Finnish conglomerate Sanitec, joining Doulton as one of their brand names. Twyford ceased manufacturing sanitary ware in the UK in 2011, resulting in all Twyford's productions being manufactured 'somewhere abroad'.

Each Christmas Thomas W Twyford gave a 60lb chest of tea to Thomas Crapper, George Crapper and Robert Marr Wharam. Edmund Sharpe is also known to have given decorated tea services, manufactured in his own pottery as a gesture of appreciation.

Porcelain Enamelled Fireclay Lavatory Range, with Overlapping Joints

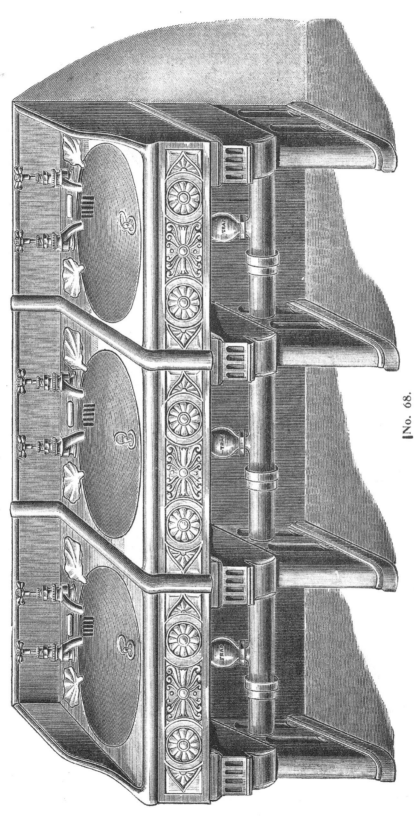

No. 68.

No. 68. Range, complete as shewn, including Three White Enamelled Fire Clay Lavatory Basins, overlapped, with Backs, Soap Trays drained into Basins, and Open Grid Overflows with Covers; Four White Enamelled Fire Clay Pedestals, and White Enamelled Fire Clay, Waste Pipe; Screw-down Brass Taps, ⅜-in., Three Hot and Three Cold; Three Brass Safety Plugs, and Three 1¼in., Lead Traps with Brass Screw Caps.

Basins, over all, 26 × 20 × 6ins.   Range, 6ft. 6ins. × 20ins.   £11 16 6 | Basins without Back 12/6 *less.*
  ,,   ,,   24 × 18 × 6ins.   ,,   6ft.   × 18ins.   £10 18 6 | With 1 Tap to each Basin 17/3 *less.*
  ,,   ,,   18 × 18 × 5ins.   ,,   4ft. 6ins. × 18ins.   £9 7 0 |

,, 69. Range complete, including Three Yellow Glazed Fire Clay Lavatory Basins without Backs, Plain Fronts, overlapped, with Soap Trays drained into Basins, and Open Grid overflows with Covers; Four Yellow Glazed Fire Clay Pedestals, Plain, Six Screw-down Brass Taps, ⅜-in., Three Hot and Three Cold; Three Brass Safety Plugs, and Yellow Glazed Channel (with Holes for Waste) instead of Waste Pipe. (Traps not included.)

Basins, over all, 18 × 18 × 5ins.   Range, 4ft. 6in. × 18ins.   £6 19 6   Basins with Backs, 10/6 *extra.*

Quotations given for Ranges containing any Number of Basins.

The gift of tea or a tea service by Messrs Twyford and Sharpe was a quintessential English gentleman's tradition of showing gratitude and appreciation, quite unlike today's perverse tradition of exorbitant bonuses for bankers, paid to them whether making a profit or loss!

Sanitary Air=tight Frames and Covers.

The Covers are made to receive Tiles, Wood Blocks, Stable Bricks, etc., to match
Floor in proximity.

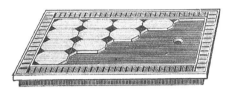

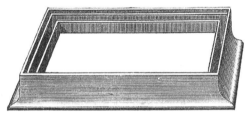

No. 503 (for Tiles).

No. 504 (for Wood Blocks, etc.)

Outside.	Opening.	Painted.	Galvanized.	Outside.	Opening.	Painted.	Galvanized.
26 × 26 in.	21 × 21 in.	£0 19 6	£1 12 6	26 × 26 in.	21 × 21 in.	£1 6 0	£2 4 0
25½ × 21½	20 × 16	0 14 6	1 3 9	25½ × 21½	20 × 16	0 19 6	1 12 6
20 × 20	15 × 15	0 10 9	0 17 9	20 × 20	15 × 15	0 14 3	1 4 0

Ventilated
Sanitary Frames and Covers.

The Covers being Ventilated are well adapted for use in Parks or Grounds connected
with Country Mansions.

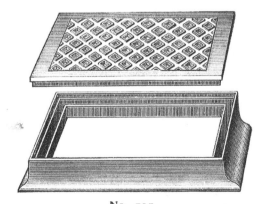

No. 505.

No. 505.	Outside		Opening			Painted.	Galvanized.
	26 × 26 in.		21 × 21 in.	...		£0 19 6	£1 12 6
,,	25½ × 21½		,,	20 × 16	...	0 14 6	1 3 9
,,	20 × 20		,,	15 × 15	...	0 10 9	0 17 9
,,	26½ × 11½		,,	21 × 6	...	0 8 9	0 14 9

Thomas Crapper and Frederick Humpherson

Thomas Crapper & Co and Humpherson & Co had much in common, both were family run businesses established during the 19th century in Chelsea, London. Thomas Crapper and Frederick Humpherson were decisive leaders who had the ability to anticipate what the public would want and buy. Their strategy was how they could best fulfill demand ahead of their competitors. They were pragmatic businessmen and employers with a strong urge to improve not only their own personal circumstances but further possessed a genuine desire to provide improvements in sanitation to the general public at large via their standing as responsible sanitary engineers.

The Humpherson family connection with Chelsea predates that of the Crappers. Research has revealed that Frederick Humpherson's grandfather, James Humpherson, was the first to live and work in Chelsea during the 1830s. James John Humpherson was a coachsmith by trade, originally from Finsbury, London. He migrated to Chelsea c1832 with his wife Henrietta and their young family including Edward Humpherson, making their home in Sidney Terrace. Tragically James Humpherson died of dropsy in 1836 age 34 years. The local undertaker Mr Butt arranged the funeral and James Humpherson was interred on the 6th April 1836 in St Luke's churchyard, Chelsea. Charles Dickens married Catherine Hogarth at St Luke's on 2nd April 1836.

Edward Humpherson returned to Chelsea from Kent having married Matilda Farmer in 1853. Edward set up home at 7 John's Place, Off John Street, near Marlborough Road, he was a Master Carpenter employing one man. Edward and Matilda had six children, one died an infant, leaving four sons and one daughter. Edward's sons Frederick, William, Alfred and Charles all became plumbers. Edward Humpherson moved home in 1862, just a short distance to 45 Marlborough Road, he is recorded in Simpson's 1863 Trade Directory, occupation: carpenter. Edward suffered a major setback on 12th January 1863, *'On Friday morning between the hours of 2 and 3am, a destructive fire broke out on the premises of Mr Edward Humpherson, Marlborough Road, Chelsea. Very considerable damage was done.'* (Chelmsford Chronicle 16th January 1863)

Frederick Humpherson was an intelligent young man who, it seems, learnt much from London's premier Master Plumber. Frederick had an inventive mind, similar to his mentor Thomas Crapper, which blossomed on completion of his plumbing apprenticeship, served with T Crapper & Co between 1871-75.

Butlers' Pantry Sinks.

No. 137.

20 × 14 × 7 in. deep outside		...	£0	16	9
23 × 16 × 7	,,	...	1	2	6
24 × 17 × 7	,,	...	1	5	6
28 × 17 × 7	,,	...	1	8	6
30 × 20 × 7	,,	...	1	12	0

Waste Hole $1\frac{5}{8}$ in   Overflow 2 × 1 in.

No. 137.

Housemaids' Sinks.

No. 138.

28 × 18 × $10\frac{1}{2}$ in. deep outside	...	£2	3	6
30 × 20 × $10\frac{1}{2}$	,,	...	2	5	6
32 × 18 × $10\frac{1}{2}$	,,	...	2	5	6

Waste Hole 2 ins.

These Sinks are made with or without Overflow.

No. 138.

Kitchen Sinks.

No. 139.

24 × 18 × 6 in. deep outside	£1	5	0
30 × 20 × 7	,,	1	8	6
36 × 20 × 7	,,	1	11	6
36 × 24 × 7	,,	1	17	6
42 × 24 × $8\frac{1}{4}$	,,	2	15	0
48 × 24 × $8\frac{1}{2}$	,,	3	7	6

Waste Hole $2\frac{1}{2}$ in.

No. 139.

Prices quoted do not include Washer and Plug.

Incredibly, after only one year working as a journeyman plumber Frederick established F Humpherson & Co, plumbers & decorators, 331 King's Road, Chelsea. Frederick's business partners were his father Edward, who provided the premises and decorating skills and his brother William, who was also a newly qualified plumber. The partnership was, however, short lived being dissolved on 21st October 1876 by mutual consent. The plumbing business was carried on by Frederick and William as Humpherson & Co. This new partnership was dissolved six years later, in 1882.

Following the 1876 partnership dissolution, Edward Humpherson continued trading on his own account as a carpenter and builder from 420 King's Road, Chelsea. It was from this address that he was declared bankrupt on the 28th January 1885. (London Gazette) Edward recovered and continued to trade as a builder and decorator working into his 80s from 474 Upper King's Road, Chelsea, where in 1913 he passed away.

William Humpherson gave up plumbing and relocated to Devon in 1882, where, after working for a ceramic tile manufacturer, he and William Hexter, previously employed as a foreman at the Great Western Pottery, Devon established in 1889 the firm of Hexter, Humpherson & Co, Potteries, Newton Abbot, Devon, manufacturers of bricks, tiles and drainage pipes.

William Hexter was born at Plymouth, at 9 years of age he was an inmate of Wolborough Union Workhouse. At his death in 1922 he left a large estate valued at £32,000.

According to the Gloucester Citizen 19th September 1894 'Messrs Hexter, Humpherson & Co, Potteries, Brick & Tile Works, Kingsteignton, near Newton Abbot, Devon were completely destroyed by fire. The buildings covered an acre and a half, the damage estimated at £50,000. The loss was covered by insurance, however 250 hands were left idle for some time.'

In 1897 Hexter, Humpherson acquired the Aller Vale Pottery (established 1865) and in 1901 they acquired the Watcombe Pottery in Torquay, amalgamating the two companies as The Royal Aller Vale & Watcombe Pottery, later simplified to Royal Watcombe, trading until 1962.

William Humpherson enjoyed a very successful and financially rewarding business career. William died on the 1st May 1939 and left £26,884. His son Ernest continued the pottery manufacturing business. Hexter, Humpherson & Co Ltd was taken over in 1964, the brick and pipework operation closed in 1968 and the company ceased trading in 1970.

Closet Basins.

Cottage Flush=down.

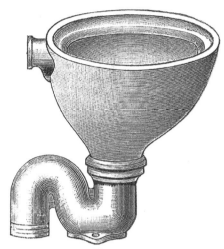

No. 204.

"The Washout."

No. 205.

No. 204.	Cane and White	...	4/3
	Cane and Printed ...	...	5/6
	Plain White	...	7/9
	White and Printed	...	9/3

No. 205.	Cane and White	...	12/-
	Cane and Printed...	...	15/-
	Plain White	...	14/-
	White and Printed	...	15/-

Slop Receivers.

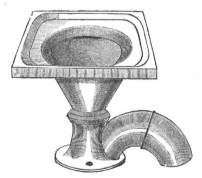

No. 206.

No. 207.

No. 206. White Ware with Fixed Top and
Flushing Rim ... £1 10 6
White and Printed ... 1 13 6

No. 207.	Cane and White ...	...	8/9
	Plain White	...	10/9
	White and Printed	...	12/9

If with Supply Arm and Flushing
Rim ... Extra 2/-

Frederick Humpherson, trading as Humpherson & Co, successfully continued to grow his plumbing business and subsequent to the departure of William he took on his younger brother Alfred, who had also recently trained as a plumber with T Crapper & Co. Frederick focused his attention on inventing and manufacturing, once again following in the footsteps of Thomas Crapper. Further, like Crapper, Frederick described himself as a sanitary engineer which, in Victorian England, conferred status in the title.

In 1885 Frederick Humpherson was granted two patents, one for an improved syphon cistern and the other for pipe joints. Further patents followed: in 1886 improved stop cocks, 1888 a siphonic flushing cistern, 1891 improvements in pedestal water closets and 1892 a patent for automatic or intermittent siphonic flushing apparatus. Frederick Humpherson won several awards at various sanitary exhibitions for his inventions, which of course raised his company's profile and broadened his clientele.

By 1887 Humpherson & Co, patentees and manufacturers traded from the Beaufort Works, 297 Fulham Road, South Kensington, London. The company's catalogue of Sanitary Earthenware Appliances for Builders, Plumbers and Brass Founders, advertised Humpherson's own manufactured products e.g. valve closets etc, also included was the popular Patent Wash Out Closet in various designs, a product which Humpherson obtained from James Woodward & Rowley of Swadlincote, Derbyshire, a long established manufacturer (established 1790) who also supplied items to Thomas Crapper & Co.

Humpherson & Co continued to expand, opening new showroom premises in 1902. Frederick Humpherson, head of Humpherson & Co, died in 1919 age 65 leaving a widow and a daughter. Frederick passed on his business to his brother Alfred whose sons Sidney and Ernest were also plumbers. Alfred Humpherson passed away in 1945 bequeathing his business to his son Sidney and daughter Edith. By the 1970s Humphersons expanded to incorporate 7 kitchen and bathroom showrooms in South East England. Humpherson & Co was sold in 1981 to the owners of Siematic Kitchens. No members of the founding family are now involved in the original firm of Humpherson & Co.

Edward Humpherson's youngest son Charles, born 1863, having worked for his father and brother as a plumber, left London c1905 for Brighton, Sussex where his wife Ellen ran a guest house. Charles and his son Reginald also a plumber operated independently as C Humpherson & Son, plumbers,

Improved Independent Boilers,

With Boiler to Level of Grate Bars.

Smoke Flue and Sockets can be fixed at the right, left, or back of Boiler, as required, but they will be fixed at the back unless ordered otherwise.

Also with Waterway extended to the bottom. Prices on application.

PRICES and Sizes as follow :—

No. 279.

Height over all.		Diameter of Boiler.	Fitted with one Flow and Return Sockets.	Approximate Heating Power. 2 in. Pipe.	Price.		
Ft.	In.	In.	1, 1¼ or 1½ in. Wrt. Sckts.	Ft.	£	s.	d.
2	8	13	2 in. Cast Sockets	300	4	15	0
2	8	13	2 ,,	300	5	2	6
3	0	13	2 ,,	350	5	12	6
3	6	13	2 ,,	400	6	0	0
3	6	15	2 ,,	500	6	15	0
4	0	15	2 ,,	600	7	17	6
4	0	18	2 ,,	900	10	15	0
4	6	18	2 ,,	1050	11	17	6
5	0	18	2 ,,	1200	13	0	0
5	0	21	2 ,,	1400	15	5	0
5	6	21	2 ,,	1600	16	5	0
6	0	21	2 ,,	1700	17	10	0
5	6	24	2 ,,	1800	21	7	6

Wrought Welded Conical Boilers.

Suitable for Offices, Coach-houses, Small Greenhouses, Conservatories, Show-rooms, &c., and for General Amateur Purposes.

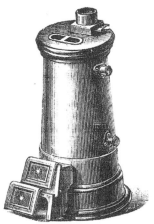

No. 280.

PRICES. including Flow and Return Sockets and Damper, as per drawing.

External Sizes.				Heating Power.			
Height 21 in.	Diameter 10 in.	2 in. Pipes 150 ft.	...	£2	7	6
,, 26	,, 12	,, 200	...	2	15	0
,, 29	,, 14	,, 300	...	3	2	6
,, 36	,, 14	,, 400	...	4	5	0
,, 34	,, 16	,, 500	...	5	0	0
,, 38	,, 16	,, 600	...	5	15	0
,, 42	,, 16	,, 700	...	6	10	0

Two decided advantages are gained by making this small Boiler conical, as shewn above—

First: The flame and heated gases impinge upon the side of the Boiler with greater effect.

Second: The possibility of fuel sticking fast in the Boiler instead of falling down as it burns is obviated.

Brighton and so continued the Humpherson family plumbing tradition. By coincidence Thomas Crapper once owned a house in Brighton, at 21 Powis Square.

No piece on Humphersons would be considered complete without mention of the claim ... who designed the Pedestal Wash Down Closet? Over many years, since at least 1960, there have been several publications, often including unfounded claims, counterclaims and even retractions on the subject. It has been claimed, and generally accepted, that Frederick Humpherson's flush down pan and trap closet The Beaufort, awarded a certificate of merit by The Sanitary Institute of Great Britain when exhibited at Leicester in 1885, was The Original Wash Down Closet. The basis for this claim appears to rely on an illustration within Humpherson & Co's 1890 catalogue for The Beaufort Pedestal Closet. Other claims are that Frederick Humpherson designed the world's very first one-piece wash down WC pan in 1884, known as The Beaufort and coined the term 'flush down or wash down closet'.

There is no tangible evidence that Frederick Humpherson held a patent or registered title for either a flush down or wash down pedestal closet during the 1880s or indeed earlier than 1891. The Author's research has produced evidence that in 1879 well known and respected plumber, sanitary engineer, patentee and author William Paton Buchan of Glasgow, patented a wash down closet, known and registered as the 'Carmichael Wash Down Accessible Closet'. Buchan's wash down closet was available in one-piece earthenware, either S or P trap outlet, pedestal design and could be fixed freestanding i.e. exposed. In 1879 he had it made with a full box rim flush and by 1882 offered it with a jet flush via a spreader, aimed at the American market, where he also held the patent.

Buchan's patent Carmichael Wash Down Accessible Closet is illustrated in W P Buchan's book *Plumbing* published in 1883. In addition Glenn Brown, in his book *Water-Closets* dated 1884, reproduced from Brown's articles in The American Architects and Building News during 1883, illustrate Buchan's Carmichael Wash Down Closet invented in 1879 and the claim *'met with success in England.'* Indeed Buchan in his own book gives details of many locations, his illustrious clients and institutions where his Carmichael Wash Down Closets had been installed with satisfaction.

P J Davies, in his book *Practical Plumbing* (1885), refers to Buchan's Patent Carmichael Wash-down Closet, as the ACME of hopper closets.

The Patent "Triune" Closet
(Earthenware Basin and Stand, and Lead Trap).

Front View of Plain Basin
and Stand.

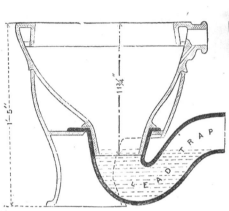

Section through Basin,
Stand and Trap.

Front View of Basin and
Stand, with Raised
Ornamentation.

Side View of Plain Basin.

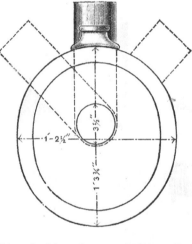

Plan looking down, shewing the
different Angles at which
Trap may be fixed.

Side View of Plain Stand.
NOTE— Dotted line shews the Shelf in
Stand.

No. 160. Earthenware Basin and Stand, Plain White or Ivory, with P Lead Trap				£1 16	0
,, ,, with Raised Ornamentation			,,	1 18	0
,, ,, Plain White or Ivory, and Iron Stand			,,	1 14	0
,, ,, with Raised Ornamentation			,,	1 15	0

If with S Trap   ...   .   Extra   2/6

  ,,   Basin Printed inside   ...   ...   ,,   2/-

Plain Basin   Tinted outside   ...   ...   ,,   4/-

Ornamented Basin   ,,   ,, ...   ,,   14/-

Whilst the Author is aware of other worthy claims to the invention and title of the wash down WC e.g. John Roe's Pedestal Wash Down Closet 1878, Edmund Sharpe's Patent Flushing Rim Closet 1855 and Stephen Green's one-piece Pedestal Syphon Closet 1852, the aforementioned must surely end any further claim on behalf of Frederick Humpherson to the wash down water closet in any form.

Setting this aside, Frederick Humpherson must have been delighted when his patent Triune Wash Down Closet of 1891, manufactured on his behalf by James Woodward & Rowley, was promoted and included in Thomas Crapper & Co's 1895 catalogue. Certain proof that Thomas Crapper continued to support and encourage his former protégé, although it must be acknowledged that, Frederick Humpherson's achievements, substantial as they were, never attained the fame and fortune of The Master –Thomas Crapper.

Thomas Crapper advertisement 1908

11

THE WHARAMS

Following the death of his beloved Maria in 1902, Thomas Crapper age 68 and in declining health lost his appetite for work, he knew it was time to move over and leave the field to others. Thomas Crapper retired in 1904, he was head of the firm that bore his name for 43 years. Thomas planned and engineered a smooth transition of ownership for his highly successful business, taking the difficult, but pragmatic decision, not to leave his majority share holding to his nephew George Crapper. Thomas had recently incorporated his business into a limited company with a few selected shareholders, he promoted George from foundry manager to director, he regarded George almost as his own son, providing him with a substantial minority share in the new company as reward for his loyalty.

Thomas was aware his much younger, junior partner Robert Marr Wharam wanted the business, importantly Wharam via his wealthy father-in-law had access to sufficient funds to buy a majority stake in T Crapper & Co Ltd. A deal was negotiated which allowed Thomas to make future financial provision, not for just one, but several members of his family, allowing himself the freedom of a less stressful retirement with time to indulge in his passion for growing flowers and fruit. Thomas was an active member of The Royal Horticultural Society.

Thomas Crapper died on the 27th January 1910, he provided substantial legacies for several of his nephews and nieces, but would never know if his decision to hand his firm to Wharam would be vindicated. The Wharam family successfully operated Thomas Crapper & Co Ltd for the next 59 years. In 1963 it fell to a Wharam to decide the future of Thomas Crapper's beloved business.

The census for 1841 Thorne Quay enumerates Joseph Wharam age 20, a merchant's clerk, residing three doors away from the Crapper household. Joseph was the son of Richard Wharam a farmer, formerly a mariner, of

THOMAS CRAPPER & CO., LTD.,

SHOWROOMS AND OFFICES:
120 KING'S ROAD, CHELSEA, LONDON, S.W.3.

By Appointment to H.M. The King.

Works: DRAYCOTT AVENUE, CHELSEA, S.W.

Telephone: Kensington 3711 (two lines). Telegrams: "Crapper, Chelsea"

"MARLBORO" LAVATORIES.

No. 1610.

No. 1523.

SILENT VALVELESS WATER WASTE PREVENTERS. Will flush when two-thirds full.

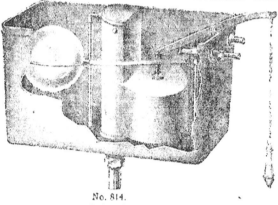

No. 814.

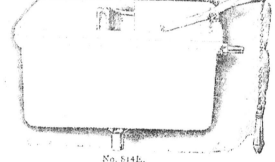

No. 814b.

Painted, Galvanized, Porcelain Enamelled, Earthenware and Lead-lined Cases.

THE "KENWHAR."

THE "IMPROVED MARLBORO."

With Lead Trap

No. 1318.

No. 568.

COMPLETE ILLUSTRATED CATALOGUE AND PRICES ON APPLICATION.

Advertisement from Architects' Compendium 1919

Fishlake, Yorkshire. Joseph's first wife Harriett died in 1845 leaving him with a baby son to care for. Joseph Wharam married his second wife Ann Marr at Thorne, Yorkshire in 1847. In 1851, Joseph, aged 30, a ship owner's clerk, lived next door to Captain Charles Crapper and family including a young Thomas Crapper. Thorne's steam packet boat trade rapidly deteriorated causing Joseph Wharam to relocate his family c1859 from Thorne to Rotherham where he became a railway navigation agent. Unfortunately this line of work failed, by 1870 Joseph and his family moved to London, residing in Shaftesbury Court, Kensington. Joseph established a coal merchant business which he ran until his death age 88 in 1906.

Robert Marr Wharam born 1853 at Thorne, Yorkshire was age 6 or 7 when his family moved to Rotherham. Thomas Crapper was already living in London by the time Robert was born. Thomas Crapper's parents lived in London between 1860-73, they would have much in common with Joseph Wharam – perhaps his firm delivered coal to the Crappers or he needed a plumber, he would certainly recognise their familiar Yorkshire name when he came across it. A probable scenario is that their paths crossed again in London where they resided in the same vicinity.

It appears young Robert Marr Wharam gained employment as a commercial clerk with Thomas Crapper in or about 1870, a position he held when he married Amelia Richmond in 1883. Amelia was the daughter of a wealthy London grocer. Frederick Humpherson's apprenticeship indenture dated 8th April 1871 was signed by Frederick, his father Edward and Thomas Crapper, the document was witnessed by Robert Marr Wharam who was barely a year older than Frederick Humpherson.

Around 1887 Thomas Crapper promoted Robert Marr Wharam to junior partner, an astute move which was of mutual benefit to both parties. Robert Marr Wharam was a strict Methodist lay preacher and refrained from the usual vices of alcohol and smoking, apparently he had an aloof disposition. Thomas was confident he could trust Wharam to manage his office and financial affairs. In fact, Wharam embraced his new role and surprisingly, in 1891, published a small book entitled *Hints on Sanitary Fittings and their Application*. Wharam's book contains just 31 pages and is illustrated with Thomas Crapper's wares, at about this time he referred to himself as a 'Sanitary Engineer'.

Robert M Wharam's book was primarily intended for *'the benefit of owners of house property etc., not for those sufficiently conversant with the*

THOMAS CRAPPER & Co., Ltd.

Manufacturing Sanitary Engineers.

120, KING'S ROAD, CHELSEA, S.W.3.

Telephone

BATHS, LAVATORIES, CLOSETS, SINKS.

PLUMBERS' BRASS WORK &c.

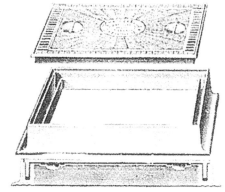

No. 496

L.C.C. SOIL PIPES & FITTINGS.

Cast Iron Drain Pipes and Fittings.

Deep Seal Manhole Covers & Frames.

Also

Double Seal or Double Cover

Manhole Covers.

New Illustrated Catalogue (1928) sent on receipt of Trade Card.

The Plumber and Decorator Journal 1928

By appointment to the late King George V

THOMAS CRAPPER & CO. LTD.

BUILDERS' MERCHANTS & MANUFACTURERS OF SANITARY EQUIPMENT

BATHROOM AND SANITARY FIXTURES SUPPLIED
TO THE NORMAN HOTEL, ST. LEONARDS-ON-SEA

The recently Enlarged and Improved Showrooms at these
Premises Display an Attractive Selection of Bathroom
Fitments of Various Design and Colour, including several
Fully Equipped Bathrooms

A VISIT OF INSPECTION WOULD BE HIGHLY ESTEEMED

"MARLBORO WORKS"
120 KING'S ROAD, CHELSEA, LONDON, S.W.3

TELEPHONE. KENS. 4831

Hastings and St Leonards Observer 20 May 1939

principles and methods in this little book to render it superfluous.' An obvious criticism of Wharam's book is nowhere within its pages of enlightenment and advice does it suggest or recommend the house owner employ the services of a fully qualified, experienced, professional plumber.

A further surprise was the discovery that in 1897 George Crapper and Robert Marr Wharam were granted a patent No.724 for 'Improvements in or relating to Automatic Syphon Flushing Tanks'. The invention was for a 'water waste preventer, siphon-discharge'. An air valve fitted to the crown of the siphon is opened by a float, as the water level rises to activate the flush automatically. The invention didn't prove to be successful.

Thomas Crapper was renowned for being a good employer, hence many of his employees worked for him their entire working lives. Upon Thomas Crapper's retirement in 1904 the Wharam family became the majority shareholders in Thomas Crapper & Co Ltd. The Chelsea Directory for 1905 describes T Crapper & Co Ltd as *'Brass Founders and Engineers, Manf. of Sanitary Appliances, Heating Apparatus, Electrical, Hydraulic, Steam & Gas Ftgs., Lead, Zinc, Glass, Colour and Varnish Merchants.'*

Thomas Crapper's legacy was, for now, in safe hands with the Wharam family. The new owners respected the old firm's history and reputation, they nevertheless wished to inject fresh ideas whilst retaining the best of the rest. In 1905 Thomas Crapper & Co Ltd negotiated to build a new showroom and warehouse on the Fulham Road, however the deal fell through, instead the firm acquired a lease on a distinctive, prestigious, large 3-storey Victorian house and grounds, previously used as an auction house and formerly let to a physician at 120 King's Road, Chelsea. The company converted the property to create exclusive showrooms with offices, ultimately purchasing the freehold in 1922.

Thomas Crapper & Co Ltd were featured in an article in the Plumbers' Journal on the 1st October 1914; *'The firm do not intend to follow the example of many manufacturers and employers of labour who have taken the despondent view of the present National crisis and plunged themselves into unwarranted hysteria of false economy by reducing staff, cutting down on wages and withdrawing advertisements etc.'* In fact Crapper & Co issued a new catalogue and the article glowingly noted *'exquisitely printed replete with illustrations of many new patterns of sanitary appliances.'*

BY APPOINTMENT

SANITARY ENGINEERS
TO THE LATE KING GEORGE V.

Thomas Crapper
& Co., Ltd.

120, KING'S ROAD,
CHELSEA - S.W.3.

KENSINGTON 4831 (4 lines)

Merchants of Sanitary Equipment

A varied selection of white and coloured bathroom and sanitary fixtures is displayed in several model bathrooms in our showroom, also various types of modern kitchen fittings.

Large stocks of lead, cast iron and alloy rainwater and soil goods and all types of plumbers' fittings.

Advertisement from Coronation 1953
Chelsea Souvenir Programme

Crapper's new King's Road showroom featured all the latest appliances including working displays of water closets. The Kenwhar closet with Crapper's 814 cistern was reportedly demonstrated clearing several pieces of paper in 4 seconds, the firm boasted the same closet with a 3 gallon cistern cleared 10 pieces of paper with just half a flush in 6¼ seconds!

Robert Marr Wharam established the firm's motto *'A place for everything and everything in its place.'*

Following the death of George Crapper in 1916, the Wharam family became the sole owners of Thomas Crapper & Co Ltd. Robert Marr Wharam and George Crapper must have had a good working and personal relationship to the extent R M Wharam was appointed an executor in the will of George Crapper.

Robert Marr Wharam was elected president of The Royal Warrant Holders Association for the year 1925. Wharam's elevated position also promoted his company's name, products and service to the elite of the business world, his high office necessarily requiring his attendance at many formal and social functions engaging with wealthier members of society.

Chelsea, and in particular the old Marlborough Road area, was changing fast in appearance, Marlborough Road was renamed Draycott Avenue in 1907. The long established local tradesmen who lived over the shop were cleared away, replaced by upmarket residential flats. Crapper's historic Marlboro' Works still appeared on a map of Chelsea dated 1935 however it was sold for redevelopment and demolished soon thereafter.

Robert M Wharam, with his son Robert G Wharam, wisely reinvested the financial compensation received for their old Marlboro' Works by carrying out major reconstruction and substantial extensions to 120 King's Road, Chelsea. It is evident from an advertisement placed in the Hastings and St Leonards Observer on the 20th May 1939 that the firm's owners deliberately retained the old foundry name, "Marlboro Works", when constructing new manufacturing works to the rear of their King's Road premises.

In July 1939 a complimentary trade press report stated *'being holders of the Royal Warrant the gilded Coat of Arms still remains on the front of the premises. The principle features of the alterations, the ground floor has been lowered to pavement level and the side drive into the Marlboro' Works has been built over to provide several floors of stock rooms. The whole of the basement has been lowered several feet. An additional entrance has been*

Earthenware Urinals.

Flat Wall Lipped Bedford
With Perforated Rim and Spreader.

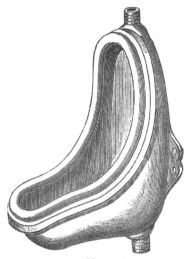

No. 264.

	Large	Medium	Small
Height	18 in.	16 in.	14 in.
Diameter	16 × 13 in.	13 × 11 in.	12 × 11 in.
	23/6	19/9	15/6

Angular Lipped Bedford
With Perforated Rim.

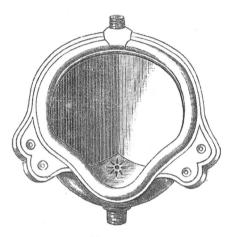

No. 265.

Height	12 in.	11 in.	10 in.
Diameter	18 × 18 in.	16 × 16 in.	14 × 14 in.
Length of Side	12 in.	11 in.	9 in.
	23/6	19/9	15/6

Angular Bedford
With Perforated Rim.

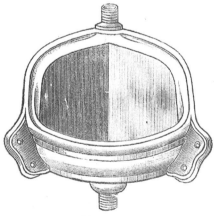

No. 266.

	Large	Medium	Small
Height	12 in.	11 in.	10 in.
Diameter	18½ × 15 in.	16 × 13 in.	14 × 12 in.
Length of Side	12 in.	11 in.	9 in.
	15/6	13/-	10/6

Angular, with Lip,
Perforated Spreader.

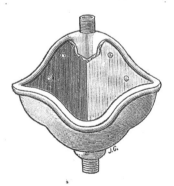

No. 267.

Height 10½ in.;   Diameter 13 × 13 in.

Length of Side 9½ in.

11/6

provided, trade customers use the old entrance and a handsome new entrance has been made for visitors to the showrooms, a new oak staircase to the main displays on the first floor. In the centre of the building is a new office for the Managing Director Mr R G Wharam M.R.San.I. from which he is able to supervise and control all departments and advise upon any matters that may be referred to him.' The article concludes *'We congratulate the veteran chairman of the company Mr R M Wharam whom we have known for more years than we care to remember, the present Managing Director and managers and staff of this old established firm upon the reconstructions.'*

Robert Gillingham Wharam was appointed a director of Thomas Crapper & Co Ltd in 1904, he gradually assumed greater responsibility for the running of the business, ultimately inheriting the company in 1942 on the death of his father Robert Marr Wharam age 89. The Wharams gradually decreased manufacturing their own goods in preference to buying in 'own brand' and other manufacturers' branded products. The company moved with the times, offering traditional and modern styles, in particular art deco and coloured sanitary ware as they became popular. T Crapper & Co Ltd survived two World Wars, greatly helped by successfully trading on its long established reputation for quality and service, whilst maintaining the Royal Warrants originally bestowed on Thomas Crapper.

For several years during the 1950s Robert G Wharam, Chairman of Thomas Crapper & Co Ltd, divided his time between London and South Africa. In his absence his managing director Robson Barrett held delegated responsibilities for effectively running the business. Robert G Wharam, like Thomas Crapper before him, had no direct male heir to inherit the business. In 1963 age 78, his health failing, he sought offers for the company from rival firms. Having obtained certain assurances regarding staff contracts he made the decision to sell Thomas Crapper & Co Ltd to John Bolding & Co (established 1822) of London, who were Crapper's main competitor.

Robert Gillingham Wharam died four years later age 82, he had a distinguished life and career. Robert G Wharam was born in Fulham, London in 1885, he served his country during WW1 and was awarded the UK Naval Medal and British War Medal. Wharam was in The Royal Navy and The Royal Naval Air Force, he was designated AC1 Air Control Officer 1st Class, he transferred to the Royal Air Force during February 1918. It is possible Robert G Wharam was involved with The Thornycroft Seaplane Lighter H21, a vessel

Earthenware Urinals.

Flat Wall
(No Supply)

No. 268.

Angular
(No Supply)

No. 269.

	Large	Small	
Height	8 in.	8 in.	$17 \times 12\frac{1}{2} \times 6\frac{1}{2}$ in. high.
Diameter	12×8 in.	$11 \times 5\frac{1}{2}$ in.	Length of Side ... $11\frac{3}{4}$ in.
	6/6	4/9	4/9

Automatic Flushing Cisterns for Urinals.

The Supply Valve having a Regulating Tap can be adjusted so that the contents
of Cisterns may be discharged as required.

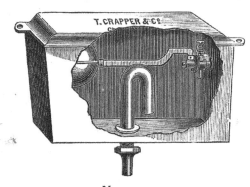

No. 270.

No. 270. Painted Cast Iron, 1-gallon 12/6

Ditto   ditto $1\frac{1}{2}$,, ... 15/6

Ditto   ditto   2 ,, ... 18/6

If Galvanized (either Cast or Wrought Iron)   Extra   4/6

towed behind a frigate or destroyer prior to launching. Flying was in its infancy and aircraft carriers were still in their early development. In 1965 R G Wharam became an Honorary Freeman of The Royal Borough of Kensington & Chelsea, having undertaken outstanding work for civil defence throughout the Second World War. He served on Chelsea Council for 19 years, was Deputy Mayor 1940-41, elected Mayor 1941-44 and Deputy Mayor again 1944-46. He became an Alderman of The Metropolitan Borough of Chelsea in 1948 and was a long serving Member of the Royal Sanitary Institute.

Robert Gillingham Wharam, owner of Thomas Crapper & Co was also President of Chelsea Football Club, a position currently held by Roman Abramovich. Chelsea Football Club is a hugely successful world famous football team who were once known as 'The Pensioners'. One can only image the fame they might have enjoyed had they been known as 'The Crappers'!

SANITARY APPLIANCES

MANUFACTURED BY

JOHN BOLDING & SONS, Ltd.,

Grosvenor Works, Davies Street, London, W.

Telegraphic
Address:

"BOLDINGS
LONDON."

⁜

Telephones:
5077 Gerrard.
818 Mayfair.

View of Showrooms, Davies Street, W.

View of Branch, 298, Euston Road, N.W.

BATHS.

LAVATORIES.

PEDESTAL, VALVE, AND

SYPHONIC CLOSETS.

URINALS.

SINKS.

Head Offices and Factory.

C.I. DRAIN PIPES

AND FITTINGS.

L.C.C. SOIL PIPES.

AIR-TIGHT COVERS.

FIRE HYDRANTS.

PUMPS.

Established
A.D. 1822.

View of Marble Works, 10 & 12, Euston Buildings,
N.W.

View of Foundry, 8, Eden Street, N.W.

Illustration from John Bolding & Sons Ltd Catalogue No.7 c1910

12
CHAPTER

THEY THINK IT'S ALL OVER

The Wharams long association with Thomas Crapper & Co, almost 100 years, came to a close when, in 1963, Robert G Wharam sold T Crapper & Co Ltd to John Bolding & Sons Ltd. Boldings, established in 1822, were Crapper's main rival in the supply and distribution of sanitary ware and plumbing equipment throughout the London area and beyond. In 1961 John Bolding employed approximately 280 people. Wharam, it is understood, secured an assurance that T Crapper's staff service contracts would be honoured, regrettably this did not happen in every instance.

Robson Joseph Barrett commenced work at T Crapper & Co in 1904 age 14, he progressed in the company from counter hand, sanitary ware salesman, clerk and ultimately managing director, a position he held until 1963. Soon after Boldings' acquisition of Crappers, Barrett's contract was terminated.

Robson Barrett gave almost sixty years service to T Crapper & Co, apart from WW1 service 1914-18, when he volunteered and joined the Royal Horse & Royal Field Artillery Regiment. Barrett saw active service in France and Italy. In September 1918, he was seriously wounded resulting in 30% disability and was granted a pension of 12/- per week for just eighteen months. Barrett was awarded the Military Medal for bravery in the field, entitling him to use the letters M.M. after his name.

Boldings restructured Crapper & Co in a manner which, at that time, was viewed as ruthless, whereas today it is the expected behaviour of city slickers. Between 1963-66 the King's Road became the trendy location for fashionable shops, forcing property prices to escalate. On the 27th August 1966 Boldings closed the Marlboro' Works, King's Road. Crapper's premises were sold and the business was integrated into Messrs Boldings' premises at Davies Street. The amalgamated business suffered severe adverse trading conditions during 1969 resulting in both John Bolding & Sons Ltd and T Crapper & Co Ltd being placed in administration.

Urinal Range
With Automatic Flushing Cistern.

No. 260.

No. 260.   Range of 3 Semi-circular Urinals in Strong Enamelled Fire Clay, 42 in. high, with Enamelled Slate Facings, Automatic Cistern with Brackets and Copper Flushing Pipes, complete **£25 12 6**

If with St. Ann's Marble in lieu of Slate Facings

Extra £3 5 0

Quotations given for Ranges with any number of Partitions.

Just when 'they think it's all over' T Crapper & Co Ltd was acquired by an overseas private equity investor, specialising and targeting insolvent turnaround acquisitions, to buy, rationalise and sell on. The John Bolding Group of companies comprised of several integrated businesses including Thomas Crapper, all were eventually sold on with the exception of Thomas Crapper & Co Ltd which was retained in a dormant status for the next thirty years. Simon Kirby, an architectural antiques dealer and sanitary ware enthusiast tracked down the whereabouts of Crapper & Co and persuaded the owner to sell him the company including Crapper's iconic brand name. Simon, with other enthusiastic investors, relaunched the famous old company in 1999. The phoenix of Thomas Crapper & Co continues successfully today, based not in cosmopolitan Chelsea, but in the tranquil Shakespearean surroundings of Stratford-on-Avon, Warwickshire. All Thomas Crapper's unique bathroom products are manufactured in the UK, maintaining the quality and service which Thomas Crapper was renowned for.

The saddest day in the long distinguished history of T Crapper & Co undoubtedly came in 1966 when the old firm vacated 120 King's Road, Chelsea. Waste skip upon waste skip were filled with T Crapper & Co business and employment records, outdated catalogues, photographs that once proudly adorned the staircase, showroom and boardroom were discarded, no longer valued or wanted. Discontinued sanitary ware, original patterns, old wooden moulds, nothing was retained that had no obvious immediate value. Reportedly one or two lucky souvenir hunters managed to salvage a few items from the skips, regrettably in just a few hours of vandalism 105 years of historic Crapper company records and memorabilia were discarded, destroyed and lost forever. Legend has it that four of Thomas Crapper's Royal Warrants were rescued, hopefully one day they will be reunited with Thomas Crapper & Company Ltd.

Recently a similar fate was almost repeated when Messrs Twyfords, once the world's largest sanitary ware producer ceased manufacturing at their Alsager premises in South Cheshire. Fortunately on that occasion boardroom paintings and photographs of distinguished past chairmen including Mr T W Twyford together with leaded stained glass windows that once adorned prestigious offices at Cliffe Vale were saved for posterity by Simon Kirby of Thomas Crapper & Co Ltd. In 2013 Twyford's old Bath Street factory, long abandoned, was also demolished, very little now remains in the UK of this

Soil Pipe Terminals.

No. 423.

No. 424.

	3 in.	3½ in.	4 in.
No. 423. Lead ...	5/9	6/6	7/6

	2 in.	2½ in.	3 in.	3½ in.	4 in.
No. 424. Lead	2/2	2/9	3/4	3/10	4/4

No. 425.

No. 426.

	4 in.			3 in.	3½ in.	4 in.
No. 425. Lead	6/6	No. 426. Lead ...	6/6	7/6	8/6

once great firm except for the frontage to their former Cliffe Vale Pottery, converted to flats during 2008.

The great walnut and leather personal office desk used for many years by Thomas William Twyford now resides at the Stratford-on-Avon premises of Thomas Crapper & Co Ltd, it is used daily by Thomas Crapper chairman Simon Kirby, I'm sure Mr Twyford would approve!

The memory of Thomas Crapper is perpetuated in a variety of ways, including:

- The 27th January is officially celebrated as Thomas Crapper Day. It is the day that the founder of Thomas Crapper & Co passed away.
- A portrait miniature of Thomas Crapper, painted by Edith Bertha Crapper (1892-1979) grand niece of Thomas Crapper, was bequeathed to The Victoria & Albert Museum National Portrait Gallery, London. The V&A website incorrectly, at the time of writing, refers to Miss E B Crapper as the daughter of Thomas Crapper.
- The London Borough of Bromley placed a blue plaque at No.12 Thornsett Road, Bromley commemorating Thomas Crapper who lived there from 1895 until 1910.
- There are 4 surviving T Crapper & Co cast iron inspection covers in situ at Westminster Abbey, London, reportedly very popular with visitors. There are a number of Thomas Crapper manhole covers and gulley grids remaining in the gardens of Sandringham House. The greatest number of known Crapper manhole covers is located at Oakham, Rutland, where 11 examples survive.
- Several publications and numerous websites feature Thomas Crapper. *Flushed with Pride* (1969) by Wallace Reyburn was the first biography of Thomas Crapper. Reprinted in 2011 with a foreword and epilogue by Simon Kirby.
- St Lawrence's Parish Church, Hatfield, near Thorne, West Riding, Yorkshire installed a stained glass window, known as The Millennium Window, which incorporates a small tribute to Thomas Crapper. A panel in the centre of the window depicts a silhouette of a water closet pan. The Millennium Window celebrates significant people and events from the locality during the past 1000 years.

Rawlings' Patent Blowpipe.

Can be worked in any position in a strong wind.

No. 389.

It gives a Flame 10 in. long, which can be immediately regulated to a flame smaller than that obtained from a candle by simply rotating Plug of Cock.

No. 389.   ...   15/-

Gas Force Pump.

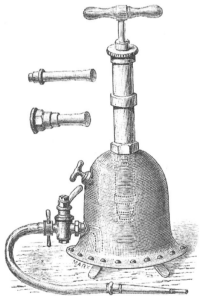

No. 390.

No. 390. Complete with Rubber Tube and Brass Connection   £3 5 0

If with Copper Cylinder extra   7/6

Briscoe's Patent Water and Gas Plug.

A most useful Tool for closing the ends of Pipes, Unions or Taps of any kind, from ¾ in. to ½ in.

No. 391.

No. 391. As shewn   ...   3/3

- Thomas Crapper's grave has been restored twice, most recently in 2002 by Thomas Crapper & Co Ltd.
- Thomas Crapper the racehorse owned by APIS.uk.com sponsored by Thomas Crapper & Co.
- In 2013 a touring play entitled *Royal Flush*, the story of Thomas Crapper performed by the Rich Seam Theatre Company was described as *'A revealing, touching and hilarious account of the man who became Britain's number one at getting rid of number two!'*

Speaking Tube Fittings.

Round Mouthpieces and Whistles
(Screwed Iron Pipe Gauge).

No. 359.

		$\frac{1}{2}$	$\frac{5}{8}$	$\frac{3}{4}$	1 in.
Cocus Wood	each	1/6	1/9	2/-	2/3
Ebony, or Box Wood	,,	1/9	2/-	2/2	2/6
Ivory	... ,,	4/11	6/-	6/3	6/9
Brass	... ,,	4/-	4/6	5/-	6/-

Oval Mouthpieces and Whistles
(Screwed Iron Pipe Gauge).

No. 360.

		$\frac{1}{2}$	$\frac{5}{8}$	$\frac{3}{4}$	1 in.
Cocus Wood	each	2/1	2/6	2/9	3/-
Ebony, or Box Wood	,,	2/4	2/7	3/-	3/4
Ivory	... ,,	7/5	8/-	8/7	9/9

Round Mouthpieces and Whistles
(Screwed inside to take Flexible Braided Tubing).

No. 361.

		$\frac{5}{8}$	$\frac{3}{4}$	$\frac{7}{8}$	1 in.
Cocus Wood	each	1/7	1/10	2/1	2/4
Ebony, or Box Wood	,,	1/10	2/1	2/3	2/7
Ivory	... ,,	5/6	6/2	6/8	7/6

Oval Mouthpieces and Whistles
(Screwed inside to take Flexible Braided Tubing).

No. 362.

		$\frac{5}{8}$	$\frac{3}{4}$	$\frac{7}{8}$	1 in.
Cocus Wood	each	2/2	2/6	2/10	3/1
Ebony, or Box Wood	,,	2/6	2/10	3/2	3/5
Ivory	... ,,	8/-	8/9	9/7	10/6

$\frac{5}{8}$ *in. Mouthpieces screwed* $\frac{1}{2}$ *in. gauge.*

Red Ivory Indicators   ...   Extra 6d. each.

13

CHAPTER

... AND FINALLY

The following pages contain extracts from a recently discovered supplementary document secured within a copy of Wallace Reyburn's book *Flushed with Pride, The Story of Thomas Crapper.* This book is inscribed *'To Edith'* - Edith Crapper was a spinster great niece of Thomas Crapper. The document originated from Frank Crapper, great nephew of Thomas Crapper and cousin of Edith.

Frank Crapper was in his 86th year when he wrote the document and gave the book to Edith, age 77. Wallace Reyburn described Edith in his book as *'the last to bear the name of Crapper of those directly related to the firm'* clearly overlooking Frank Crapper, who upon reading Reyburn's book penned his own supplement for the benefit of Edith Crapper.

Frank Crapper, born 1884, commenced his career as a plumber with Crappers, subsequently becoming a successful surveyor and estate agent. Frank's father, Charles Crapper 1852-1912 was a beneficiary under Thomas Crapper's will. Frank was a beneficiary in the wills of his spinster aunts Emma and Maria Crapper, who between them had inherited Thomas Crapper's home, possessions and a share of his money upon his death.

Edith Bertha Crapper was a professional artist specialising in miniature paintings and illustrations. Edith's father, George Crapper 1854-1916, was also a beneficiary under Thomas Crapper's will, he was formerly a director of Thomas Crapper & Co Ltd.

The Wallace Reyburn book, referred to above, still containing Frank Crapper's original document is now in the possession of Simon Kirby of Thomas Crapper & Co Ltd.

Edith Bertha Crapper passed away in 1979 age 87.

Frank Crapper passed away in 1983 age 99.

Garden Fittings.

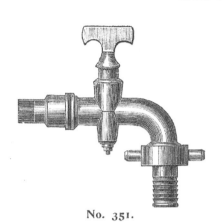

No. 351.

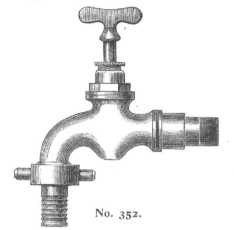

No. 352.

	$\frac{1}{2}$ in.	$\frac{3}{4}$ in.	1 in.
No. 351. Screw Bottom Bib Cock with Union on Nose ...	4/6	6/3	11/3
,, 352. Screw-down ditto ditto ditto ...	4/6	6/3	11/3

No. 353.

No. 354.

	$\frac{1}{2}$ in.	$\frac{5}{8}$ in.	$\frac{3}{4}$ in.
No. 353. Branch Pipe with Cock, Rose and Union ...	4/6	5/6	6/6
,, 354. Jet and Spreader for ditto	1/9	2/3	2/9

No. 355.

No. 356.

No. 357.

No. 355. Double Union

$\frac{1}{2}$	$\frac{3}{4}$	1 in.
1/7	2/3	3/2

No. 356. Jet

$\frac{1}{2}$	$\frac{5}{8}$	$\frac{3}{4}$ in.
8d.	10d.	1/-

No. 357. Rose

$1\frac{1}{2}$	$1\frac{3}{4}$	2	$2\frac{1}{2}$	3 in.
1/4	1/7	1/10	2/2	2/6

No. 358.

	$\frac{1}{2}$ in.	$\frac{5}{8}$ in.	$\frac{3}{4}$ in.	$\frac{7}{8}$ in.	1 in.	
One-ply Garden Hose	$4\frac{1}{4}$d.	$5\frac{1}{4}$d.	$6\frac{1}{2}$d.	$7\frac{1}{2}$d.	$8\frac{3}{4}$d.	per foot.
Two-ply ditto	$5\frac{1}{4}$d.	$6\frac{1}{2}$d.	$7\frac{1}{2}$d.	$8\frac{1}{2}$d.	$10\frac{1}{2}$d.	,,
Three-ply ditto	$6\frac{1}{4}$d.	$7\frac{1}{2}$d.	$8\frac{3}{4}$d.	$10\frac{1}{2}$d.	$1/0\frac{1}{4}$	,,

A SUPPLEMENT TO THE STORY OF THOMAS CRAPPER

'I knew him well and had great admiration for him, I often stayed with them and when Aunt died he had me to stay several weeks until Emma was installed. He was always very kind to me but apart from this quality of kindness, now I come to think of it, he also had loyalty, he was definitely loyal to old Wharam. ... the only time I remember him saying anything against him; he was annoyed that Wharam was pushing the sale of other brass-work instead of their own. ...

On one of my visits he had just had a new peach-house installed, I remember how pleased he was to tell me that he offered to lend the glazier a hand with the glass, they had not gone very far before the glazier said, "You carry on Sir, I will soften the putty for you."

He always carried a gold matchbox. On a tour of inspection of the work in progress at Sandringham the Prince of Wales asked him for a match to light his cigar, he did not smoke, he had not any matches, (too late the matchbox).

My earliest recollection of him was at Brighton when they were living at 21, Powis Square. My memory is a bit dim of these times but I do certainly recollect the old Cook stuffing me into the lift cage and winding me up from the Basement to the Ground Floor but this was put a stop to by the sudden disappearance of the Cook, I seem to remember something about the gin decanter being diluted with water. He was always very jolly with us kids, I do remember he was especially jolly on the evening of our arrival, he had brought me down from town, I gather there had been a celebration lunch in town.

Mrs. Finch, the owner of the "Finborough Arms", Finborough Road, Kensington, (now owned by Brewers), told me she used to often open a bottle of champagne in the morning for him and his brother George (my grandfather), not an unusual custom in those days I believe. ...

He also told me a tale about his Father who went to help a woman who was being beaten up by a man and when he was getting the better of the fight the woman turned on him and shouted "You leave my husband alone" and together they threw him off the pier into the sea. "Remember Frank," he said, "never interfere between husband and wife." ...

A short while ago the cistern in the outside toilet of my house had to be renewed and when I came to look what make of cistern had been fixed to my

Telegrams, " CRAPPER, CHELSEA."   Telephone No. (01769) 450 522.

THOMAS CRAPPER & CO. LTD.,

PATENTEES AND MANUFACTURERS

.. OF ..

Sanitary Appliances.

ENGINEERS BY APPOINTMENT TO

Their Late Majesties
King Edward VII AND King George V.

OFFICES AND STORES :
THE STABLE YARD, ALSCOT PARK, STRATFORD-ON-AVON, WARWICKSHIRE.
(CV37 8BL)

ESTABLISHED 1861.
INCORPORATED AS A LIMITED COMPANY (№ 82482) IN 1904.

PRODUCERS OF THE WORLD'S MOST AUTHENTIC PERIOD-STYLE SANITARYWARE.

LATE OF THE BOROUGH OF CHELSEA;

NOW IN THE IMMORTAL BARD'S BOROUGH OF STRATFORD-ON-AVON, WARWICKSHIRE.

All goods marked with the CRAPPER name are exclusive to T.C. & Co; we NEVER simply re-label another company's products.

surprise I found just the initials "T.C. & Co." on it. ... I have nothing but praise for my Great Uncle.

The firm I was associated with at the time in 1913 were carrying out improvements at 4, St. James Square, the town house of the "Cliveden" Astors. They wanted a W.C. in their principal Guests Bathroom, the bathroom in the centre of a large rambling house had ventilation but no external wall, to comply with the byelaws a W.C. had to have an external wall and it would be impossible to obtain official consent. The firm were very anxious to please "Nancy Astor", this was before she became Lady Astor famous as the first lady to take her seat in the House of Commons. She little knew that she had an unlawful "loo". I wonder now the house is in public occupation if it is still there.

The Plumbers of the 18th and 19th century were craftsmen in lead; fine ornamental cisterns, there is one in the basement of Locks, the Royal Hatters, in their early 18th century building in St. James Street, and before the days of a regular supply of water, huge lead lined storage cisterns had to be installed in large houses. One of these was done away with and the tank room converted into the Nursery Bathroom during the work at 4, St. James Square, tons of old lead to credit against the cost of the new Bathroom.

Without in any way minimising the research that has gone into the preparation of this story, I feel he would like to record the debt he owed to his brother, George, much older than himself, for the help he gave him in his early days when he joined him in London, and as an acknowledgement of this help and proof of his loyalty in his will he left all his money to the children of this brother.

F. Crapper
17, Elm Avenue,
W.5.
February, 1970'

CPSIA information can be obtained at www.ICGtesting.com
Printed in the USA
LVOW02*1147290914

3718LVUK00005B/6/P